WITTGENSTEINIAN VALUES

Wittgensteinian Values

Philosophy, religious belief and descriptivist methodology

EMYR VAUGHAN THOMAS

Routledge
Taylor & Francis Group
LONDON AND NEW YORK

First published 2001 by Ashgate Publishing

Reissued 2018 by Routledge
2 Park Square, Milton Park, Abingdon, Oxon, OX14 4RN
711 Third Avenue, New York, NY 10017, USA

Routledge is an imprint of the Taylor & Francis Group, an informa business

A Library of Congress record exists under LC control number: 00131672

ISBN 13: 978-1-138-71634-6 (hbk)
ISBN 13: 978-1-315-19703-6 (ebk)

Contents

Acknowledgements

Along the years many people have contributed to my attaining the stage of intellectual development and outlook that has enabled me, through the grace of God, to produce this work. Geoffrey Hines, O.G. Roberts (a former headmaster who instilled a passion for learning in so many of his pupils, including myself) and Professor Max Wilcox (the latter formerly my undergraduate tutor at the University of Wales) deserve special mention. I also owe a great deal to my parents for the sacrifices made and encouragement given when providing me with an education.

I am also grateful to Dr Oswald Hanfling for the patient and thorough tutorial support given in the past, and to the Rev. Dr Cyril Barrett (his encouraging comments on an early post-graduate project of mine were an important motivating factor in my setting out to develop my thoughts on the central themes of this work).

Very special thanks are due to Professor Stuart Brown of the Open University. I have benefitted an immense amount from his intellectual example and from his kind support over the years.

Very special thanks are also due to Professor Paul Helm of Kings College London. His penetrating skill as a research supervisor has been the fundamental inspiration that has led to the translation of my often vague ideas into something more coherent. Though I feel that my stumbling efforts have not done full justice to his guidance, I do hope that this inadequate expression of thanks will at least go some way towards making up the deficit.

Special gratitude is also due to my wife, Susan, for giving my life a sense of completeness without which no enterprise of this kind could have been sustained.

Thanks are due to the the editor of *International Philosophical Quarterly* for permission to reproduce material from my paper 'Wittgensteinian Methodology and Religious Belief' which appeared in the journal in September 1999.

I am grateful to Cambridge University Press for permission to use material previously published in my papers 'Wolgast on Innocence', *Philosophy* 69 (1994) and 'Wittgensteinian Perspectives (Sub Specie Aeternitatis)', *Religious Studies* 31 (1995).

Finally, thanks are due to Blackwell Publishers for permission to use material from my article 'From Detachment to Immersion: Wittgenstein and the "Problem of Life"', *Ratio* 12 (1999).

List of Abbreviations

The number in brackets after each entry refers to the section number in the bibliography where full bibliographic details can be obtained.

AE	Holland, R.F., *Against Empiricism* (1).
AEMIP	Holland, R.F., 'Absolute Ethics, Mathematics and the Impossibility of Politics' (1).
ARG	Sutherland, Stewart, *Atheism and the Rejection of God* (1).
BH	Hebblethwaite, Peter, *Bernanos* (2).
C	Tolstoy, Leo, *A Confession and Other Religious Writings* (3).
CL	Emerson, R.W., *The Conduct of Life* (3).
CP	Phillips, D.Z., *The Concept of Prayer* (1).
CV	Wittgenstein, Ludwig, *Culture and Value* (1).
DC	Bernanos, Georges, *Dialogues des Carmelite* (2).
DE	Rilke, R.M., *Duino Elegies* (3).
DI	Phillips, D.Z., *Death and Immortality* (1).
EA	Winch, Peter, *Ethics and Action* (1).
EC	Feuerbach, L., *The Essence of Christianity* (3).
FA	Phillips, D.Z., 'Ffydd Athronydd – Eglurdeb neu Atebion?' (1).
FAF	Phillips, D.Z., *Faith After Foundationalism* (1).
FATM	Dilman, Ilham, *Freud and the Mind* (1).
FPE	Phillips, D.Z., *Faith and Philosophical Enquiry* (1).
GE	Gaita, Raimond, *Good and Evil: An Absolute Conception* (1).
GJB	Sutherland, Stewart, *God, Jesus and Belief* (1).
HAH	Hammelmann, H.A., *Hugo von Hofmannsthal* (3).
HHP	Hamburger, M., ed., *Hugo von Hofmannsthal Poems and Verse Plays* (3).
IE	Phillips, D.Z., *Interventions in Ethics* (1).
ISS	Winch, Peter, *The Idea of a Social Science* (1).
J	Bernanos, Georges, *Journal d'un cure de campagne*, translated by Pamela Morris (2).
LE	Wittgenstein, Ludwig, 'Lecture on Ethics' (1).
LHS	Dilman, Ilham, *Love and Human Separateness* (1).
LLC	Hofmannsthal, Hugo von, 'The Letter of Lord Chandos' (3).
LRB	Wittgenstein, Ludwig, 'Lecture on Religious Belief' (1).
LSLW	Shields, P.R., *Logic and Sin in the Writings of Ludwig Wittgenstein* (1).
LWDG	Monk, Ray, *Ludwig Wittgenstein: The Duty of Genius* (1).

MN	Phillips, D.Z., 'My Neighbour and My Neighbours' (1).
MRB	Shooman, A.P., *The Metaphysics of Religious Belief* (1).
NB	Wittgenstein, Ludwig, *Notebooks 1914–16* (1).
NHM	Bernanos, Georges, *Nouvelle Histoire de Mouchette* (2).
OC	Wittgenstein, Ludwig, *On Certainty* (1).
PI	Wittgenstein, Ludwig, *Philosopical Investigations* (1).
PM	Sharp, F.M., *The Poet's Madness: A Reading of Georg Trakl* (3).
POR	Tilghman, B.R., *An Introduction to the Philosophy of Religion* (1).
PPL	Dilman, Ihlam, *Philosophy and the Philosophic Life* (1).
RBLG	Phillips, D.Z., 'Religious Belief and Language-Games' (1).
RFGB	Wittgenstein, Ludwig, 'Remarks on Frazer's "Golden Bough"' (1).
RMRB	Brodsky, Patricia Pollock, *Rainer Maria Rilke* (3).
RMRG	Graff, W.L., *Rainer Maria Rilke: Creative Anguish of a Modern Poet* (3).
RMRH	Heerikhuizen, F.W. van, *Rainer Maria Rilke* (3).
RMRSW	Rilke, R.M., *Selected Works* (3).
RPW	Dilman, Ilham, 'Reason, Passion and the Will' (1).
RST	Phillips, D.Z., *R.S. Thomas: Poet of the Hidden God* (1).
RWE	Phillips, D.Z., *Religion Without Explanation* (1).
RWECW	Emerson, R.W., *The Collected Works of Ralph Waldo Emerson* (3).
SG	Alvarez, A., *The Savage God* (3).
SILAR	Dilman, Ilham, *Studies in Language and Reason* (1).
SKPC	Dilman, Ilham, 'Self-Knowledge and the Possibility of Change' (1).
SOS	Taylor, Charles, *Sources of the Self* (1).
SRFD	Thomas, Emyr Vaughan, 'D.Z. Phillips, Self-Renunciation and the Finality of Death' (1).
SS	Bernanos, Georges, *Sous le soleil de Satan* (2).
SW	Winch, Peter, *Simone Weil: The Just Balance* (1).
TAW	Kerr, Fergus, *Theology After Wittgenstein* (1).
TLP	Wittgenstein, Ludwig, *Tractatus Logico-Philosophicus* (1).
TS	James, Henry, *The Turn of the Screw* (1).
TTMS	Winch, Peter, *Trying to Make Sense* (1).
UPS	Winch, Peter, 'Understanding a Primitive Society' (1).
WA	Rhees, Rush, *Without Answers* (1).
WAP	Tolstoy, Leo, *War and Peace* (3).
WAR	Phillips, D.Z., *Wittgenstein and Religion* (1).
WAZ	Canfield, John, 'Wittgenstein and Zen' (1).

WEA	Tilghman, B.R., *Wittgenstein, Ethics and Aesthetics* (1).
WOTS	Dilman, Ilham, 'Wittgenstein on the Soul' (1).
WPRB	Coughlan, Michael, 'Wittgensteinian Philosophy and Religious Belief' (1).
WRPV	Malcolm, Norman, *Wittgenstein: A Religious Point of View* (1).
WWR	Schopenhauer, Arthur, *The World As Will and As Representation.* (Volume number indicated by capital Roman numerals in between the abbreviation and the page number) (3).

1 Self-Renouncing Religious Belief: The Wittgensteinian Position

Introduction

The aim of this work is to contribute to extending the Wittgensteinian descriptivist treatment of religious and ethico-religious values. It seeks to expose and ultimately weaken certain tendencies of thought which have emerged within Wittgensteinianism itself. Such tendencies comprise implicit presuppositions about the nature of the self-renouncing dimension which the Wittgensteinian approach has shown to be important to understanding the character of religious belief. In other words, this work recognizes the importance of the concept of self-renunciation in helping us to understand the nature of religious belief. What it takes issue with is the way the notion of a self-renouncing faith has often been treated by philosophers who subscribe to a Wittgensteinian approach. Presuppositions about self-renunciation can, indeed have, resulted in distorted philosophical accounts of the nature of religious belief. It is hoped that this work will go some way towards minimizing the future capacity for perpetuating such (often unconscious) distortions.

D.Z. Phillips and Ilham Dilman have explicitly sought to elucidate the character of religious belief by reference to the concept of self-renunciation. It is my contention that these two philosophers do not in this respect represent an engagement in an isolated or eccentric enterprise cut off from the mainstream of Wittgensteinian philosophy. Rather the implicit attempt at the ascription of a self-renouncing character to religious belief (by which I mean belief in God as found in the Christian traditions) is more widespread. In some complex ways it interlinks with some central facets both of Wittgenstein's own work and that of a far wider body of his followers who subscribe to a Wittgensteinian approach.

It is important to try to clarify what is meant by the phrase 'the Wittgensteinian approach'. I am not wanting to claim there is a single, discrete approach linked by a clear, explicit creed of orthodoxy. The influence of Wittgenstein has been very wide in philosophical circles. There is a loose grouping of philosophers engaged in continuing what they see as a distinctive approach to philosophy originated by Wittgenstein. Though this grouping of thinkers does not necessarily yield a clear, comprehensive, systematic and integrated view on the nature of religious belief as self-renouncing.

However, certain implicit themes or elements occur widely in Wittgensteinian writings. This first chapter will seek to outline five dominant traits of this kind. I shall refer to these as 'elements' or (sometimes, for the sake of variation) 'theses'. The former term is probably the more accurate in the sense that what is at issue is generally not a matter of strictly formulated propositions but rather underlying and unformulated assumptions. The five elements are: the Absoluteness-Element; the Anti-Consolation Element; the Unreflectiveness-Element; the Perspective-Element and the Anti-Metaphysical Element.

In what follows I shall call a view that has all, or most of the above elements,'the Wittgensteinian position'. This will provide a usage that marks something a little more specific than 'the Wittgensteinian approach', as defined above.

It is important to appreciate that I do not insist that every Wittgensteinian writer discussed in Chapter 1 necessarily subscribes *explicitly* to each of the five elements around which I structure the discussion. That may be true of some. However, the key issue is that these five elements link together and set the parameters for what genuine, self-renouncing faith, is taken to be. Indeed, the interrelation between these elements is so close that if one were to subscribe to one then there would probably be a certain intellectual 'pull' towards embracing many, if not most, of the others. It is no part of my purpose to give a listing of just who falls into this temptation. My role is more the therapeutic one of trying to free us from the attractive 'force' of this temptation so that we can pursue the descriptivist aim, freeing ourselves of as many presuppositions as possible.

It is worth anticipating one important potential objection at the outset. An uncritical defender of Wittgensteinian writings might insist that philosophers such as – for example – D.Z. Phillips, Ilham Dilman or Peter Winch – *have* no presuppositions but simply attempt to describe religious life as it is. They, like Wittgenstein himself, are faithful to the latter's injunction to look, to understand and to describe.

However, this objection has only a superficial force. Despite their emphasis on the practice of a descriptive approach sensitive to *variety*, what we all too often find in the work of some of the most vociferous Wittgensteinians is a remarkably uniform underlying view of just what genuine religion is. For example, D.Z. Phillips' *From Fantasy to Faith* (London: Macmillan, 1991) contains an apparent wealth of examples from a range of writers. However, on closer examination, the apparent variety dissolves into a largely uniform logical model comprising most of the elements I referred to above. Religious belief seems to invariably be something held as absolute, unreflective and unrelated to any metaphysical dimension. The full force of these points will be clearer by the time we re-visit this issue in Chapter 4.

A further objection that might be made to the legitimacy of my purpose is that I imply an essentialism to those writers who I shall criticise which is really a misunderstanding of their descriptivist purpose. For, it might be insisted, a descriptivist is concerned with *instances* and not *essences* in the sense of some sort of *core* to religious belief.

Two points can be made in reply to this objection. Firstly, any philosopher, even the most thoroughgoing descriptivist, is concerned with degrees of generalisation. Someone that only studies instances is not a philosopher but a scholar of religious studies or whatever. The real challenge is not that of avoiding generalising but of avoiding the wrong sort of generalisation.

Secondly, if we can recognise in the latter point a legitimate area for an appropriate degree of generalisation we can also acknowledge a latitude for illicit generalisation. How we construe this needs careful handling. Generalising at the level in question tends to depart from descriptivism when it imposes a model on our thinking that leads us to assume that certain phenomena have an application far beyond the range to which they normally apply. (For example, when it assumes that some of the 'elements' to be described in the further sections of this chapter can be taken to apply beyond their natural scope). Wittgenstein was all too conscious of the threat that our thinking can be polluted by misleading analogies which we might be drawn into taking too far. This is a recurrent danger which no descriptivist philosopher should consider him or herself insulated from.

The uncritical defender of the Wittgensteinian position might insist at this point that it is no defect in a descriptive approach if it fails to encompass everything 'in the field' and that, therefore, some of the things I shall draw attention to in this book, should not be taken to imply any defect in the Wittgensteinian position as I have defined it. Although this is a fair point, it must also be borne in mind that generalisation tends towards illegitimacy when it leads us to exclude some elements which might have a legitimate place in self-renouncing faith. For example, D.Z. Phillips at many points in his extensive writings make use of a distinction between authentic religion and what he calls superstition. The latter category seems to include any and every form of metaphysical conception of religious concepts such as 'God'. This *might* constitute a method that excludes recognition of phenomena that a genuine sensitivity to variety *should* encompass. By the end of this book we will perhaps be in a better position to decide on this issue.

It is hoped that this study will inspire further meditation on the way a descriptivist methodology can be practised, as well as an on-going appropriate degree of self-examination by its practitioners – including the present author. For it is no part of the aim of this book to seek to preach from a position of superiority. Wittgenstein's writings lead us to appreciate both how difficult clarity is to achieve and to hold on to. It is in that spirit that the present work is intended to be understood.

After examining five key elements in the Wittgensteinian position in this chapter (where my main purpose will be exposition, although some criticism will also be made at appropriate points), Chapter 2 will consider an alternative conception of self-renouncing faith which is incompatible with the model of religious belief encapsulated in these five theses. This will be interpreted as an indication that the Wittgensteinian position is unable to encompass what is genuinely a self-renouncing stance. In Chapter 3 the presuppositions that have led the Wittgensteinian position to this unsatisfactory point will be explored by tracing the five constituent elements to their *Sitz im Leben*, the situation and way of life that inspired them. This will enable us to discuss in the final chapter the nature of the presuppositions inherent in what I have called the Wittgensteinian position and to acquire some sense of what it takes to arrive at a more durable descriptivist method. My aim, ultimately, is not to dismantle the Wittgensteinian position but in some modest way to press for its extension so that its application to religious belief will be broader. It will be able to extend to 'new territory', to engage with some forms of belief it has hitherto dismissed as inauthentically religious. More will be said about this in Chapter 4.

The Absoluteness-Element

The Wittgensteinian position holds that religious belief is an absolute. That is, it is an end in itself and as such is discontinuous with all relative ends. All relative ends have to do with the self and its machinations for its own solace and security.

One example is D.Z. Phillips's adoption of the distinction between absolute and relative judgements, derived from Wittgenstein's 'Lecture on Ethics'. A relative judgement about an action can be supported by reference to information about how that action contributes to a further end. If someone says: 'Make sure you secure that strip before you start drilling', this imperative can be supported by reference to the likely unpleasant consequences of not securing the strip. In contrast, an absolute judgement cannot be supported by reference to a further end. 'You ought to want to behave better', someone might insist. But if another replies, 'What if I don't?', then no further reasons related to further ends can be given. Phillips claims belief in God is an absolute in just this way.

> *...(I)f a man urges someone to come to God, and he asks 'What if I don't?', what more is there to say? Certainly one could not get him to believe by telling him that terrible things will happen to him...(For he would then be)...believing in the best thing for himself...(and not in God)* (RBLG 124).

The idea is that belief in God cannot logically come about through fear. For if fear is what prompts belief, then what we are describing is not belief in God at all but a policy for the welfare of the self. The situation is akin to the Glauconian question asked by Father Sergius in Winch's account of an example derived from a story by Tolstoy.

In the Father Sergius example Winch gives an account of the difference in Sergius' faith at the time of his succumbing to temptation from what it was previously. At the point of succumbing to temptation Sergius' religious life changed in character. Previously, the problem of lust was

> *...understood by him from the perspective of a genuine religious belief...(and)...it was not then a case of setting the satisfaction of his desire alongside the demands of his religion and choosing between them. The fulfilment of his religious duties was not then for him an object to be achieved* (EA 189).

At the time of succumbing to temptation, however, Sergius' faith had become a matter of 'an object to be achieved'. As such, it had become a relative end – an avenue to his own satisfaction.

Winch's account contains an implicit view of the nature of self-renunciation which underlies this understanding of there being these two distinct and discontinuous periods in Sergius's spiritual life (corresponding, respectively, to cases of genuine and corrupt forms of religious belief). When religious belief is a matter of something to be achieved Winch says that the 'Glauconian question' arises. This is the question raised in expectation of an answer as to *why* the religious life is worthwhile. In the particular example of Sergius it is found in the form: 'What does it matter?' This question is inviting '...a judgement explaining why religious purity is more important than the satisfaction of lust...'(EA 189). In asking this question Sergius '...contemplated the religious life as an object and asked what there was about it that made it worth while' (EA 190). Winch comments

> *But just as Adeimantus noted that when people commend justice, what they commend 'is not justice itself, but the respectability it brings', so Sergius found*

> *that if he tried to commend the religious life, what he was commending was not that at all, but the kudos and admiration it brought him* (EA 190).

A little further on Winch adds

> *If one tries to find in the object of contemplation that which makes it admirable, what one will in fact see is the admiration and applause which surrounds it* (EA 190).

In so valuing the religious life for the admiration it brought, Sergius '…could not help feeling pleasure in it' (EA 191). This gave Sergius' faith the character of a means to an end, namely self-fulfilment.

Winch gives us an important sense of the distinction between an absolute conception of ethics and a conception which looks at the ends in terms of desires which morality serves. The latter type of conception has enjoyed contemporary support in, for example, some works by Phillips Foot ('Morality as a System of Hypothetical Imperatives', *Philosophical Review* 81, 1972) and Bernard Williams ('Egoism and Altruism' and 'Morality and the Emotions', in his *Problems of the Self*, Cambridge: Cambridge University Press, 1973). But an absolute conception of ethics is much better able to account for the logical distinction between the way desire influences a person and the way moral claims make demands on him or her. The reality of this distinction leads even Williams to talk of altruism as a '…general disposition to regard the interests of others…as making some claim on one, and, in particular, as implying the possibility of limiting one's projects' (*Problems of the Self*, 250). To see their interests as having a claim on one is something very different from one simply having the desire to assist them.

A further example of the Absoluteness-Element is found in R.F. Holland's interpretation of Kierkegaard's idea of an eternal resolution as not involving acting for the sake of some end. 'In so far as one's doings cease to be…for the sake of some end…they no longer amount to bringing about…they approximate to taking things as they are, or indeed to suffering. And it is in suffering that…freedom is said to be greatest' (AE 79). Of course, some forms of endurance can be for an end, for a purpose which it is hoped to realise. But Holland's point relates to a form of endurance that is logically not ends-directed. Elsewhere Holland stresses how absolute goodness is 'antithetical to assertion of the self' (AEMIP 177) and how it involves a 'moral geometry which puts the doing of evil outside the agent's limit, while providing him with infinite space in which to suffer it…Absolute ethics is the ethics of forgoing' (AEMIP 182–183).

This helps us to recognise how the idea of an absolute is irreducible to the parameters of a consequentialist ethic, such as Utilitarianism. In some versions of the latter each agent counts as a unit, a unit whose pleasures or wellbeing can be aggregated along with others to arrive at the most favourable total distribution. Arriving at such a distribution may require some units to be dispensable. But if all are to be given equal treatment and some turn out dispensable then this affects the status of even those units that form part of the majority. The lack of dispensability of their needs is not something that pertains to *them*. It merely arises randomly from the way the contingencies happen to be structured in a particular distribution. Further, on such a model the notion of a sacrifice of the self can never have the same status as that in an absolute conception. A utilitarian saint would sacrifice him or herself for the best probable distribution of wellbeing. Though not all varieties of utilitarianism allow this form of sacrifice, even those that do allow it reflect only a contingent connection between sacrifice of self and what is ethical. Only if sacrifice of self leads to desirable consequences is it ethical. This is not the case with the Absoluteness-Element. Its proponents affirm that allegiance to what is absolute is intrinsically, not contingently, self-renouncing. To hold something as absolute is necessarily to assign a limit to the self's demands. Thus an absolute conception of ethics is in principle a self-curtailing conception.

For this reason, an absolute view of ethics sits uneasily with an Aristotelian view of the good life. For example, the modern Aristotelianism found in Jean Hampton's paper 'Selflessness and Loss of Self' (in eds. E.F. Paul, F.D. Miller and J. Paul, *Altruism*, Cambridge, Cambridge University Press, 1993) involves claiming that we are sometimes morally required to benefit ourselves over and against others. Development of the self is something we are obliged to give attention to. There is no such necessary obligation on an absolute construal of morality. In J.L. Stocks' words, morality '...may call on a man at any moment to surrender the most promising avenue to his own moral perfection' (*Morality and Purpose*, London: Routledge, 1969, 29).

One further illustration of the Absoluteness-Element is found in connection with Wittgensteinian accounts of the total disinterestedness ascribed to love of God. Dilman says that if love is pure then '...it must not be sustained by any reward – goodness must be its own reward' (PPL 90).

It is less than clear whether the nature of the disinterestedness found in various forms of Christian love can be neatly captured by the formula that such love should be an end in itself. For this implies a total separation from concerns for self-fulfilment. This is an idea which, as we shall see in Chapter 2, has no place in some conceptions of self-renouncing faith.

Someone might be tempted to ask the question: how does something come to be established as an absolute? The Wittgensteinian position denies that argument can ever achieve this. Rather, the absoluteness of some things, such as belief in God, arises from the nature of the phenomena which make human life what it is. Thus Norman Malcolm, towards the end of his famous paper on Anselm's Ontological Argument, considers why human beings have even formed the concept of an infinite being. He then suggests *some* of the background conditions for the emergence of this idea. Among them is *guilt*, guilt beyond all measure, and the storm in the soul that seeks out a forgiveness that is beyond all measure. D.Z Phillips has explored the *Sitz im Leben* of religious concepts in virtually all his extensive writings and found a far broader range of phenomena than mere guilt. Dilman and Winch similarly engage in a comparable form of often precise descriptive analysis. Of course, Wittgenstein himself, in his 'Lecture on Ethics', insisted that the idea of absolute value has to be traced to certain experiences in life – unless we do that we will never see it as anything but nonsensical. He also wrote: 'Life can educate us to a belief in God...sufferings of various sorts...(e)xperiences, thoughts, – life can force this concept on us' (CV 86).

The Anti-Consolation Element

The Absoluteness-Element entails that self-renouncing religious belief has no further ends. The Anti-Consolation Element identifies consolation for the self as the fundamental further end that is incompatible with self-renouncing faith.

Consolation, a term I borrow from an analysis of Simone Weil by D.Z. Phillips (CP 99), comprises psychological satisfaction or solace. On the face of it the Wittgensteinian position seems, in the literature, to provide two versions of an Anti-Consolation Element. The first version, let us call this the Weak Anti-Consolation Element, excludes the compatibility of particular forms of consolation with self-renouncing faith. A second version, the Strong Anti-Consolation Element, excludes consolation *per se*.

One form of the Weak Anti-Consolation Element entails antipathy towards consolation deriving from the believer's sense of his or her own holiness. Thus Gaita claims that while for Aristotle the pleasurable appreciation of the nobility of one's deeds was no obstacle to a person's virtue, this cannot be said of Christianity. In the latter there is supposedly an exclusion of the person's delight in his or her virtue (GE 89). Similarly, Dilman says we attribute a love of goodness to the saint – he cannot assign it to himself because 'to do so would be to give himself credit and so to receive a reward' (PPL 90).

The point here is that it is the reward element, this psychological fulfilment of pleasurable appreciation, which precludes this being virtuous. According to some conceptions of self-renunciation such pleasurable appreciation is not a matter of gloating but is close to what Peter Geach describes as 'a special emotional attitude, a kind of wonder, that a lover has towards himself as the bearer of this precious thing, love' (*The Virtues*, Cambridge: Cambridge University Press, 1977, 74). Alongside this, of course, we have to recognise a point made by Sutherland:

> *If a man's ethical intent is some form of integrity or selflessness, then this very intent can be the most insidious temptation of all...the fanatical search for purity...can secretly feed the furnaces of white-hot pride...* (GJB 123).

As Dilman points out (SILAR 118) virtue is not something one can contemplate as an *achievement*.

Another example of a 'prohibited' consolation is that experience of having secured reprieve from adverse events in the world. An example is found in Winch's emphasis on the difference between a man who repays a debt to avoid criminal proceedings against him and one who does so 'simply because he owes it, without any thought of any unpleasant consequences to him ensuing from his not paying it' (EA 182).

A further example of prohibited consolation is that which seeks to alleviate the self's sense of its limitations by getting satisfaction from a sense of the certainty of being compensated for any suffering encountered. D.Z Phillips makes extensive use of Simon Weil's writings on this point. He gives an example of someone who assumes a right to be compensated for every effort, a person who develops an absolute need for a reciprocation for every act of generosity or sacrifice. Such a person's life is wholly governed by the self-centred and consoling thought that he or she has claims everywhere (CP 69).

Another form of consolation that is excluded from being compatible with self-renouncing faith is that sense of a bond to God or to acting for Him. An example of this is to be found in Dilman's adoption of some thoughts by Simone Weil. Dilman says that if God is conceived of as somehow metaphysically present when we act, His presence would be 'something we can have in view when acting...(and)...this would change the character of our actions: they would no longer involve self-renunciation' (SILAR 115). If God is construed as present then ...(e)ven a martyr going to his death is supported by the bond he feels unites him to God' (SILAR 115). This is akin to Phillips' claims that self-renouncing faith does not go on 'hoping for a reciprocating touch' (RST 152).

From this latter point we can turn to discussing the Strong Anti-Element which implies that self-renouncing faith can endure a total lack of consolation. Dilman says that a martyr going to his death

> *...does not feel alone and abandoned. This is not to deny that his action may not be completely selfless and disinterested. It may be so, but it does not diminish*

> *his sense of self. It does not replace it with a void. Christ felt no such support when he was nailed to the cross...he did not know that God was with him* (SILAR 115).

Reference to the 'void' here implies not the absence of particular forms of consolation, but of all consolation. The self 'consents' to be nothing and is replaced by a void. The

> *...individual person is exposed to the thought that he is* nothing... *(and that there is)... nothing that he can be proud of, nothing on the support of which he can count, nothing he can look forward to or expect any solace from* (SILAR 23).

Self-renunciation is a state in which what 'goes out of a person goes one way; nothing that he anticipates eases his burden' (SILAR 117). D.Z. Phillips also talks of a state in which the believer realises that '...in the ego there is nothing whatever, no psychological element, which external circumstances could not do away with' (CP 70). Such a realisation suggests an antipathy to *any* consolation for the self. Phillips also directs us to the peasant in the poems of R.S. Thomas who is

> *...Endlessly ploughing, as though autumn*
> *Were the one season he knew.*
> *...No signals*
> *Cheer him; there is no applause*
> *...I can see his eye*
> *That expects nothing...* (RST 56).

Like the peasant, the self-renouncing believer is able '...to give up craving for a sense other than a sense of endurance...' (RST 55).

The Strong Anti-Consolation Element has several difficulties. Firstly, it is hard to reconcile it with some religious conceptions in which the self is renounced *through* an identification with Christ in suffering. Thus interpreted the message of the Crucifixion is that the degree of abandonment suffered by Christ is not required of the ordinary believer. Secondly, a consequence of this Strong Anti-Consolation Element is that it implies the virtual deletion of personality. Self-renunciation is conceived of as a stripping away, ultimately an almost literal absence, of the self. This view is rare in Christianity, perhaps only ever having gained any ground in early Syriac asceticism where it was not uncommon to live on top of columns for years on end or out in the open without shade from the sun and, like the animals, feeding on grass.

Thirdly, there are problems about the coherence of the Strong Anti-Consolation Element. Surely there has to be some intentionality whereby the believer's commitment is *for* something. And if it is *for* something, there is, minimally, some sort of sense of a *bond* which, we can concede, can be in some sense consoling. Even if that bond is not to a God conceived metaphysically, it is a sense of being part of some mission or worthwhile purpose. (Such a purpose must, I think, be explicitly conceived of – a point to be taken up later in connection with what I call the Unreflectiveness-Element). If we deny that then we will find it very difficult to distinguish some religious responses from some attempts at masochism. In this connection there is the example of St Lydwinne, a saint of the Middle Ages who is featured in a novel by J.K. Huysmans which illustrates the doctrine of vicarious substitution. St Lydwinne's body breaks out into a mass of boils and sores, a condition willingly accepted and prayed for as a means to take on some of the suffering of the world, channelling it away from others. Now this is an example in which there is a consolatory purpose to the suffering. Yet that is the very thing that makes this case intelligible as one of renunciation of self. Take away the purpose of deflecting suffering away from others and it appears as a repulsive attempt at masochism. This suggests that the Strong Anti-Consolation Element is not logically viable. The fact that there is a consolatory sense of purpose or bond does not invalidate the authenticity of faith and is necessary for the intelligibility of something as an instance of self-renouncing faith.

As it happens a number of Wittgensteinian philosophers can be clearly seen not to subscribe to the Strong Anti-Consolation Element when they acknowledge some consolatory role for religion. Tilghman talks approvingly of Bonhoeffer's 'consciousness of being borne up by a spiritual tradition that goes back for centuries and gives one a feeling of confidence and security in the face of all passing strains and stresses' (PR 194). Malcolm writes:

> *Yet there are many people, even in this technological and materialistic age, who observe religious practices – praying to God for help, asking Him for forgiveness, thanking him for the blessings of this life – and who thereby gain comfort and strength, hope and cheerfulness...Many would regard their faith as an undeserved* gift *from God. When overwhelmed by calamity, they arrive at a kind of reconciliation once they come to feel that these sufferings are God's will...* (WRPV 84).

Finally, it is important to note how Dilman's allegiance to the Strong Anti-Consolation Element is short-lived. For when he goes on to contrast a case in which a father plays with his son out of a sense of duty with one in which the father plays 'out of pure joy and pleasure' he says:

> *...the latter is more selfless than the former; in the pleasure he shares with his child he has forgotten himself* (SILAR 129).

He goes on to say of the former father that

> *...he keeps his 'ego' out at the expense of his 'power to give life to others' because he is unable to disentangle it from his passions and desires....this is (not) what self-renunciation means...a man who wants nothing for himself* may *be a person who does not know how to want anything without being greedy...* (SILAR 129).

In this latter quotation Dilman gives us a clear argument against the Strong Anti-Consolation Element.

We can extract a further argument from Dilman against the Strong Anti-Consolation Element. This is found in his comments on the words: 'You cannot really think of others unless you can enjoy life.' Of these words he says:

> *This is only true if 'enjoying life' means...'finding sense in what you are doing'. But this is in no way incompatible with self-renunciation...(T)he man who wants nothing for himself is not one who wants nothing at all, since he may fight for others...with passion. As for the man who turns the other cheek, he is not necessarily one who feels that nothing is worth defending...* (SILAR 130).

Here Dilman seems to allow the consolation of a 'bond' when he talks of 'finding sense in what you are doing'. Further, I suggest the act of turning the other cheek can involve a sense of such a consoling bond – the sense of a value that is worth defending.

It is worth recalling that the issue of whether and what form of consolation is compatible with authentic religion has a basis in Wittgenstein's own writings. There are indications that he toyed with the idea of a state of not wanting or desiring any satisfaction. Thus the entry for 29.7.16 in the *Notebooks* considers the idea that not wanting is the only good. However, Wittgenstein seems not to have adopted such a standpoint and so avoided what I have termed the Strong Anti-Consolation Element. Much later he explicitly, and without disparagement, recognised the consolation offered by Christianity:

> *The Christian religion is only for one who needs infinite help, therefore only for one who feels an infinite need. The whole planet cannot be in greater anguish than a single soul. The Christian faith...is the refuge in this ultimate anguish* (CV 46).

Wittgenstein does see some forms of consolation as incompatible with more genuine forms of religion. Thus we find him saying: 'If you offer a sacrifice and are pleased with yourself about it, both you and your sacrifice will be cursed' (CV 26). This seems not to be a rejection of all consolatory reward through participation in a religious rite. Rather what it condemns is a form of *self-congratulation.* Elsewhere Wittgenstein says such participation

> *...aims at satisfaction and achieves it. Or rather: it aims at nothing at all; we just behave this way and then feel satisfied* (RFGB 64).

Although we have here a reference to *satisfaction* to be had from participation in religious rites, this is readily compatible with the Weak Anti-Consolation Element. Wittgenstein can here be understood as excluding only consolation that is aimed *for* and being neutral about that which only occurs incidentally.

To summarise things thus far, it is clear that there is an antipathy towards consolation among Wittgensteinian philosophers, including Wittgenstein himself. Yet neither what I have called the Weak nor the Strong Anti-Consolation Element seems to provide a clear route to clarifying just how we should express this antipathy. The Strong Element fails because there are problems about its coherence and its faithfulness to the character of some forms of self-renouncing faith. The Weak Element seems to be too unstructured. It excludes *particular* forms of consolation. But without a full listing as to which are the 'culprit' consolations it is hard to state the 'thesis' in any simple and succinct way. Is there a way out of this dilemma?

The way out, I suggest, is to extrapolate from the implied distinction given by Wittgenstein in the last quotation above. The distinction is between consolation actively sought for and consolation which only arises incidently through living the religious life. On this basis we can state the Anti-Consolation Element, without being sidetracked by the ambiguities in the literature and having to refer to a strong or weak version. The Anti-Consolation Element so defined holds that belief in God as an end in itself cannot be *held* for any consolatory purposes. This latter definition allows that consolations may arise from such belief. What it excludes is the possibility that belief is motivated by the prospect of consolation.

Stated thus the Anti-Consolation Element does have a *prima facie* credibility. A view of self-renunciation as not seeking consolations for the self can account for some cases where we might not want to say that a person who has made great sacrifices is self-renouncing. For example, we might picture a relief worker in the Third World seeking the improvement of the infrastructure and the economic development of a region. Such a person may have sacrificed very much in the way of material wealth. Yet he or she may also be quite forceful and arrogant and have a sense of his or her own importance, not acknowledging the value or needs of others in contributing to the mission. The Anti-Consolation Element presented above could exclude this from being deemed self-renouncing on the grounds that the individual concerned is seeking psychological benefits from the contribution he or she is making.

A further point to highlight is the way the Anti-Consolation Element implies an over simple view of human nature which fails to give adequate due to the role of some forms of *emotion* in moral and religious life. The Anti-Consolation Element implies an unproblematic split between a motivation which is orientated to receiving consolation and one that is not. But in the case of some forms of emotion which have a place in religious and moral life, this neat division is decidedly unsubtle and unsustainable. The complex emotion of *compassion* is a case in point. Compassion typically involves an imaginative dwelling on the condition of another person, an identification with him or her and a regard for his or her good.

The value of compassion is stressed in many conceptions of self-renouncing faith. Yet compassion is not self-renouncing in a sense that implies any clear-cut distinction between directedness to consolation versus a neutral motivation for this. In some views of sainthood the issue of seeking consolation is simply irrelevant to the exercise of compassion by some individuals, simply in virtue of the kind of persons they are. In Chapter 2 the Bernanos case-study will provide one example of this. That chapter will also involve an illustration of the way another emotion, *joy*, can be sought by a religious believer without it being appropriate to classify this as seeking a consolation *in lieu* of believing in God as an end in itself.

None of this should be taken as implying that I think that the aspiration to remove all consolation from a person's adherence to religion has not been important in some traditions and sub-traditions within Christianity. Iris Murdoch points out that '...as soon as any idea is a consolation the tendency to falsify it becomes strong: hence the traditional problem of preventing the idea of God from degenerating in the believer's mind' (*The Sovereignty of Good*, London: Routledge, 1991, 57). The fear of this happening has been very genuine for many religious people. One is reminded of Luther's early spiritual struggle and discomfiture at the self-interested way Man turns to religion. There is also Edmund Gosse's haunting description of how his religious father went '...on his knees searching the corners of his conscience...and one by one every pleasure, every recreation, every trifle scraped out of the dust of past experience, was magnified into a huge offence' (*Father and Son*, Harmondsworth: Penguin, 1959, 78). However, these forms of antipathy to consolation constitute only particular reactions to the fear that consolation may pollute faith. They should not be taken as providing any necessary conditions as to what belongs to an authentic religious faith.

The Unreflectiveness-Element

The term 'unreflective' is borrowed from D.Z. Phillips' essay 'My Neighbour and My Neighbours' (IE 229–250) which is an assessment of Winch's paper 'Who is my Neighbour' (TTMS 154- 166) on the subject of the parable of the Good Samaritan. The latter is a good place to start expounding what I mean by the Unreflectiveness-Element.

Winch stresses the immediacy of the Samaritan's response. 'Nothing intervenes between the Samaritan taking in the situation and his compassionate reaction', in contrast to the Levite who 'went over and looked at him in a calculating way before passing by on the other side'(TTMS 156). The Samaritan's response involves no such intervening *calculation*. 'He responds to what he sees as the *necessity* generated by the presence of the injured man' (TTMS 157). On Winch's characterisation the response is immediate, non-calculating, not based on any intervening reflection and characterised by a *necessity*. Phillips agrees with all this and talks of the 'unreflective and immediate response' (IE 242). He also implies there is a certain necessity to this response when he says it is not '...something we can adopt at will'.

There is a direct link between the idea that nothing intervenes and unreflectiveness. Unreflectiveness is *directly reinforced* by the idea that nothing intervenes. (Immediacy and necessity add two further dimensions to it – namely speed and some sort of compulsion – which do not directly concern us here). The idea of unreflectiveness as something in which nothing intervenes between the person confronting a certain situation and reacting to it is rooted in what Wittgenstein says about primitive reactions such as towards other human beings: 'My attitude towards him is an attitude towards a soul (Eine Einstellung zur Seele). I am not of the *opinion* that he has a soul' (PI, II, iv). Here Wittgenstein is contrasting attitude with *opinion,* and, by implication, with *belief.* Some things which are commonly taken to be beliefs, such as the belief that another human being is not an automaton, are really derivative of primitive unreflective reactions which are intimately bound up with the emergence of language. The term 'human being' is a case in point. Both Winch and Phillips have claimed that primitive reactions of this sort should be appreciated before we assume too readily that many religious concepts and reactions are to be interpreted as *beliefs*, or causally inspired by them. (See Peter Winch, 'Meaning and Religious Language', in ed. S.C. Brown, *Reason and Religion*, London & Ithaca: Cornell University Press, 1977 and D.Z. Phillips 'Primitive Reactions and the Reactions of Primitives: The 1983 Marett Lecture', *Religious Studies* 22, 1986).

It is important to realise that what is being claimed is not that authentic religious beliefs *are* primitive reactions. Rather, they are unreflective in some ways comparable to primitive reactions. Comparable in that nothing intervenes by way of reflection between the situation the believer finds him or herself in, and his or her response. Consequently, religious beliefs are not akin to opinions; they are nearer to being attitudes.

Despite these similarities there are important differences between the unreflectiveness associated with religious belief and that pertaining to primitive reactions. Firstly, the unreflectiveness of primitive reactions is often somehow derivative of the connection of such reactions with the biological nature of mankind. Thus Winch says:

> *In its primitive form action is quite unreflective. Human beings, and other animate creatures, naturally react in characteristic ways to objects in their environments. They salivate in the presence of food and eat it; this already effects a rudimentary classification which doesn't have to be based on any reflection between 'food' and 'not food'* ('Introduction' to Simone Weil, *Lectures on Philosophy*, trans. H.S Price, Cambridge: Cambridge University Press, 1978).

Most of what goes on in what we call 'religious belief' or 'morality' does not have this rootedness in a biological context. As Phillips says in connection with the parable of the Good Samaritan and in criticism of Winch, this is not a natural but a *supernatural* reaction, whereby he means that it is exceptional.

Secondly, the Wittgensteinian position sees the unreflectiveness associated with religious belief as being distinct from that associated with primitive reactions in that it has to it a dimension in which the self is absent.

This latter point is so fundamental to the subject of our concern that some examples are called for in order to make clear how this is so. One good example is that of Winch's discussion of Father Sergius from Tolstoy's story. The essential point here is that when he succumbed to temptation Sergius' faith had become a matter of 'an object to be achieved'. It *had* become an avenue to his own satisfaction which, through reflection, can be weighed against another avenue to satisfaction, namely the seduction of the young girl who had been sent to him. Winch recounts how Tolstoy has Sergius ask the question: 'What does it matter?' This question is inviting '...a judgement explaining why religious purity is more important than the satisfaction of lust...'(EA 189). With the asking of this question the unreflectiveness has left Sergius' faith. And with its passing, *self* has crept in. Reflectiveness is self-affirming in that it marks a move away from the selfless immersion in the religious orientation to life. It leads to a state that seeks out and is orientated to the self's consolations.

D.Z. Phillips provides an example of authentic faith involving unreflectiveness linked to non-articulation in his discussion of prayers of confession in *The Concept of Prayer* (CP 66–72). He interprets prayers of confession in terms of a form of contemplation leading to self-knowledge. He explicitly distinguishes this from another form of confession which arrives at self-knowledge through a process of articulation. In Faulkner's novel *Requiem for a Nun*, Temple Stevens, by telling her story, begins to see an intelligible order in the events of her past. She '...begins to see the bearing which one thing has on another; her life begins to take a recognizable form' (CP 66). All this requires putting her story in an '...articulate form'. Phillips insists that this latter phenomenon is *not* what we find in the case of *religious* believers. He says:

> *If it were, reference to God would be superfluous. Temple is able to work out her salvation for herself. The believer finds the meaning of life in the worship of God. He does not, indeed, cannot, work it out for himself* (CP 66).

How are we to understand this?

Phillips is contrasting the self-affirming dimension to articulate confession from what he sees as a self-renouncing character of religious confession 'to' God. He stresses how Temple Stevens finds her own salvation through achieving a consoling sense of order in her life. She avoids despair by her own resources. But prayer to God is not like that. It does involve self-knowledge. But it is a very different form of such knowledge. It is knowledge of one's 'state of soul...the kind of person one is' (CP 67). Such knowledge of one's sinfulness brings a realisation that one cannot change what one is. In fact he even suggests that the ultimate form of such realisation is that: '...in the ego there is nothing whatever...which external circumstances could not do away with' (CP 70). The contrast here is with the self-reliant and self-affirming character of Temple Stevens' confession and the self-exposure to nothingness that accompanies the religious confession. The sense of such an exposure is implied to be something that is unreflective and immediate.

Another illustration of the way unreflectiveness is to be found in the Wittgensteinian position's comparison of religious belief to regulation by a *picture*. This is an idea that derives from Wittgenstein's 'Lecture on Religious Belief'. In Wittgenstein himself there is a strong emphasis on the self's subjugation to the religious picture – the forgoing of pleasures and the taking of risks on account of it. Phillips clarifies the way unreflectiveness and the self-renouncing dimension to the idea of religious belief as akin to a picture, coalesce (FPE 116). He gives an example in which the religious picture of the Last Judgement loses its hold and becomes superstitious because it is held as a *literal* picture. Phillips talks of the *fault* of the believer. This fault seems to consist of the believer seeking the consolation of a picture promising continued survival after death. This is what prevents it being a case of authentic faith – the focus is on the self's desire for comfort. The belief is held 'self-assertively' (FPE 262) and 'fostered by the natural desire to see loved ones again, and to make amends for wrongs committed' (FPE 262). All this is fuelled by reflective infatuation with the self's own advantage.

One line of thought that the Wittgensteinian position uses to support the idea of the unreflective character of religious belief is that centred on the distinction between belief *in* and belief *about*. Thus Phillips refers to the request that a believer give an account of what he is doing when he prays (CP 2). Phillips says that in such a context the believer is lost – he is being asked to give a non-religious account of a religious activity. This is really, he claims, a matter of providing a pseudo-epistemological theory to give religion respectability in other spheres of life. The unreflective character of religious belief in its natural context is supposed to be distorted by this kind of intervention which fails to take account of the nature of the religious activity itself.

The Wittgensteinian position seeks to emphasise the Unreflectiveness-Element through its use of perceptual rather than deliberative analogies. Religious belief is said to be akin to a form of *seeing*. It is worth pointing out that the idea of a *picture*, as it occurs in many forms in the practice of religion, is not straightforwardly visual and unconnected to reflective analysis. A picture such as Paul Gauguin's 'Where do we come from, what are we, where are we going?' has the effect of drawing us in to seek out and articulate our response to it. Grunwald's stark picture of the Crucifixion with its image of Christ as physically deformed in suffering gains part of its impact through the way it stands out in contrast from the tradition of the serene and harmonious depiction of the suffering Christ. It is a picture that forces many believers to reflect on the adequacy of their own mental picture of the event, a process that has elements of articulating differences and being explicit about the nature of how and what forms of contrast exist.

Does the fact of being an attitude or a perspective really entail a complete disconnection from processes of articulation and reflection? I think that a more credible picture is one in which having attitudes and perspectives on matters is often sustained and transformed by articulation of beliefs. These are not two sorts of phenomena wholly insulated from each other but often part of a wider process in an individual's orientation to the world. A process where there is an interconnection between them. And, importantly, none of this need necessarily affect the self-renouncing character of one's commitments. Indeed it can contribute to it – something which we shall see illustrated in Chapter 2.

To summarise, the Wittgensteinian position sees reflective articulation as self-affirming because it involves an avenue for securing a sense of psychological consolation or solace from such activity. *This latter sentence enshrines the essence and provides the definition of what I term the Unreflectiveness-Element.* The unreflective nature of such reactions is required in order for the self to be most truly absent. The implication is that if there was any reflection involved, then that would be an avenue for the self to intrude. Authentic religious reactions are required to have an automaticity so as to truly guarantee the absence of the self, to prevent the self impinging and impressing its character on actions and responses.

One interesting way of understanding the underlying paradigm to this thesis is to be found in John Canfield's attempt to draw comparisons between Wittgenstein and Zen Buddhism. Canfield believes that Wittgenstein has language and understanding bound together in a 'practice'. Such a practice is said to overlap with the Zen idea of 'just doing' something, that is, 'doing something with a mind free of ideas or concepts' (WAZ 383). The interesting point Canfield makes about such unreflective action is its absence of self. He illustrates this with reference to a Taoist passage from Chuang Tzu (which apparently is meant to illustrate the Zen ideal also). A man crossing a river by raft is hit by a drifting, empty boat. The man pays no heed. Later, when hit by another boat carrying people, he shouts and curses.

> *Earlier he faced emptiness, now he faces occupancy. If a man should succeed in making empty, and in that way wander through the world, then who could do him harm.*

The idea is meant to illustrate the unreflective 'just doing' of Zen as involving an emptiness devoid of self. Occupancy by reflection (and, in the Chuang Tzu quote, emotion) indicates a form of self-affirmation. I think that the paradigm which underlies the Unreflectiveness-Element is something akin to this Zen/Taoist one.

There is a possible objection to the idea that the Wittgensteinian position involves an Unreflectiveness-Element of this type. The objection is that there seems to be something akin to a process of reasoning highlighted by several Wittgensteinian philosophers. Dilman's essay on John Wisdom (SILAR 85–108) is actually entitled 'Reason and Religion' and is an attempt to show that there is a process of reasoning not tied to justification, as is that outlined by Wisdom in his essay 'Gods', but yet not a matter of whim or randomness. The form of reasoning Dilman highlights is like Wisdom's in that '...it changes one's apprehension...by making and severing connections' (SILAR 103) but different in that '...it brings about in the reasoner a transition to certain values which engage with religious belief that have no more than a merely verbal reality for him...it helps him find ...their sense – the sense of beliefs which constitute the framework of a particular perspective on life' (SILAR 103).

In reply to this objection, the point is that such a 'reasoning' is what brings about a *transition* to religious belief for some individuals. It is not, for the Wittgensteinian position, illustrative of the character of the believer's response once he or she had attained an authentic and self-renouncing faith. Similarly, D.Z. Phillips discusses a comparable process of reasoning and describes it as a '...kind of persuasion...a form of imaginative elucidation, something which will bring about the dawning of an aspect not previously appreciated' (FAF 89). Again, in Phillips, the context is one in which this process of reasoning operates to bring a person into a religious orientation and is not meant to illustrate the character of that orientation once arrived at.

Authentic religious belief, being unreflective, is not in principle subject to justification. Indeed, since the provision of a justification would seem to require reflection and articulation in some way, the idea of a justification is something self-affirming. It should be noted, however, that the Unreflectiveness-Element is not simply an 'anti-justification-thesis'. It is far broader than that. Though it entails that justification is self-affirming, it seems to extend that to all reflective processes.

A consequence of the Unreflectiveness-Element as outlined in this section is a certain form of unquestionability. For what is held in a way immune to reflective articulation cannot be subject to questioning without distorting the nature of the self's attachment to those deep convictions.

This immediately implies that the unreflective character of religious and other deep commitments is not subject to evaluation. Does not unquestionability entail a blind, uncritical imprisonment in one's deep belief and values?

Dilman has tried to deny this by seeking to stress the difference between unquestionability and non-assessibility. In a discussion of the idea of evaluation with respect to one's values he talks of 'self-criticism' and says:

> *A person who is self-critical has his eyes opened to the difficulties...in keeping faith with what one believes in. This does not mean necessarily that he can imagine and formulate objections to taking the various alternatives that present themselves to him or articulate their significance. There is a perfectly good sense in which...he may recognize what is objectionable, be able to anticipate his objections and so avoid the snares of temptation. Yet he may not be able to represent these to himself in advance...* (FATM 134).

Being able to attain self-criticism and an evaluatory stance with respect to one's values does not require articulation, at least not of the kind possessed by a philosopher. A simple, inarticulate person can still evaluate his values in the sense that

> *...a man...can be thoughtful and critical...although he is not a thinker. His thoughtfulness...is manifest in his life and actions... (A)lthough he is not a thinker...yet he may be receptive, open and even vulnerable to the way life tests and tries his values. Such a man does not take his values for granted and is responsible for his moral convictions; they are not just something given, part of his childhood cargo...His love or regard for them has been tried, and so he has been tried. We could say that his love has been transformed from a passive habit to an active attitude of will* (FATM 135–6).

Against Dilman it can be affirmed that making a man's life the testing point is hardly tenable. For a particular man may just by chance have had an easy life, a life untroubled by temptations or tribulations. This would be, ironically, a way in which the authenticity of a person's religious belief is subject to 'luck' in a way Bernard Williams has claimed morality often is. I think the Wittgensteinian position clearly, if unwittingly, leads in this direction. It implies a passivity of the self: the self is subsumed by a certain perspective on life which determines how the self sees everything. I believe that this is what Jurgen Habermas means when he claims that Wittgenstein himself practices '...the relentless expulsion of reflection in the name of understanding...to leave no middle ground between the coercion of deductive representation and the pathos of unmediated intuition' (*On the Logic of the Social Sciences*, trans. Shierry Weber Nicholson and Jerry A. Stark, Cambridge: Polity Press, 1988, 122).

The Perspective-Element

The Unreflectiveness-Element is a negative thesis in the sense that it tells us what authentic religious belief is not, rather than what it is. The Perspective-Element concerns the nature of that unreflectiveness: it portrays it as a perspective on the *whole* of life.

It is worth saying something about the implications this has for the traditional view of *belief* in God as a matter of belief in some way comparable to belief that a *being* exists. The Wittgensteinian position holds that belief in God is not to be understood as akin to the concept of belief used with respect to empirical entities. D.Z Phillips makes this point very clearly in the following quotation which employs examples from Wittgenstein (LRB 59–60):

> *If I say that something exists, it makes sense to think of that something ceasing to exist. But religious believers do not want to say that God might cease to exist. This is not because, as a matter of fact, they think God will exist for ever, but because it is meaningless to speak of God ceasing to exist. Again, we cannot ask*

> *of God the kinds of questions we ask of things which come to be and pass away, 'What brought him into existence?' 'When will he cease to exist?' 'He was existing yesterday, how about today?'*... (RBLG 127).

Coughlan, no doubt conscious of the above types of disparity between language involving facts and language about God, says:

> *Religious beliefs are not beliefs about statements in the sense of descriptions of reality, whether natural or supernatural... (B)elief in the existence of God is not belief in yet another fact...is not commitment to the truth of a statement of fact which might be theory-laden... Wittgenstein makes it clear that he takes religious beliefs to be fundamental determinants of the believer's way of seeing the world. Religious beliefs are not beliefs about facts which are coloured by some background theory; they* are *the background which colours the believer's view of the facts, they regulate for everything in a man's life...* (WPRB 238).

The idea of religious belief as a determinant of the believer's way of *seeing* things I take to be equivalent to the idea that religious belief is a *perspective* on things. The term 'perspective' is widely used in this connection. Thus Phillips portrays faith as a certain sort of *perspective* on life. The term 'perspective' is used in, for example, (FA 11; RWE 167; FAF 115 ff.). Winch also uses the actual term (EA 178, 179, 190), as does Dilman (SILAR 214) and Tilghman (POR 214). Other *visual* terms are also used by Wittgensteinians, such as 'seeing' (e.g. 'seeing grace in all things', RST 104) or 'way of looking' (e.g. FPE 209). Sometimes the term 'attitude' is used (e.g. FPE 209; also Winch's elucidation of Wittgenstein's idea of the ethical will as an 'attitude to the world' in Winch, EA 118). In my account of the Wittgensteinian position I shall use the one term 'perspective'.

It is worth noting that the idea of a perspective is a methodological device in the context of a descriptive account of religious belief. Its purpose is to enable us to be a little clearer as to the nature of religious belief. It is no part of the Wittgensteinian position's claim that believers have a perspective in the way they have a heart, a liver or a brain. The issue is not one about the relative merits of an existential claim but one that relates to the adequacy of the idea of a perspective to do justice to the character of religious belief as we find it.

The Perspective-Element portrays the character of religious belief as something all-encompassing and directed to an awareness of a connectedness and wholeness in the phenomena encountered in life. Coughlan says that religious belief is the background which colours the believer's view of (or perspective on) the facts (WPRB 238). By this he means *all* facts. Phillips talks of seeing '... (people and things) with the whole of existence as their background; to see them *sub specie aeternitatis*' (RST 129). This idea of seeing the world as a whole comes from Wittgenstein. In Wittgenstein's *Notebooks* the entry for 7.10.16 includes the claim that 'the good life is the world seen *sub specie aeternitatis*'. Whereas the usual way of seeing things is from the midst of them, the view *sub specie aeternitatis* is from outside. Things are seen with the *whole world* as their background. In Tractatus 6.45 it is said:

> *To view the world sub specie aeterni is to view it as a whole – a limited whole. Feeling the world as a limited whole – it is this that is mystical.*

This idea of seeing the *whole* is linked in Wittgenstein to the idea of independence from the world. How are we to understand this idea of independence of the world?

In the *Notebooks* the entry for 13.8.16 Wittgenstein discusses the possibility of being happy despite the misery of this world. 'The only life that is happy is the life that can renounce the amenities of the world. To it the amenities of the world are so many graces of fate.' There is also the notion of an independence from the vicissitudes of the world in the 'Lecture on Ethics'. This is expressed in the idea of absolute safety: 'I am safe, nothing can injure me whatever happens.'

The 11.6.16 entry in the *Notebooks* has the actual phrase 'independent of the world':

> *I cannot bend the happenings of the world to my will: I am completely powerless. I can only make myself independent of the world, and so in a certain sense master it, by renouncing any influence on happenings.*

In is significant that Wittgenstein refers here to a certain form of mastery over the world. I shall return to this point later. First, I shall emphasise the link between the view *sub specie aeternitatis* and the idea of independence of the world.

How is the idea of independence from the world linked to the idea of seeing the world as a whole, the view *sub specie aeternitatis*? The following comments enable us to see one way in which these notions are linked.

The context of *Tractatus* 6.45 is that of the problem of life. This is the problem of 'the sense of the world' (TLP 6.41), the world in which the subject is completely powerless (NB 11.6.16) and, in no causal sense, able to be independent of fate, to determine happenings. The solution to that problem is seen '...in the vanishing of the problem' (TLP 6.521). The implication of *Tractatus* 6.522 is that it is 'the mystical' which secures the vanishing of the problem. And, as we have already noted, *Tractatus* 6.45 equates the mystical with 'feeling the world as a limited whole' and with, in turn, the ability to view the world *sub specie aeterni*. The difference this form of viewing makes is to enable the subject to live in the present. For '...the man...fulfilling the purpose of existence...no longer needs to have any purpose except to live' (NB 6.7.16). This is what it means to be '...living in eternity and not in time' (NB 6.7.16). Such a way of living enables the self to be independent of fate, to have no fear, even in the face of death (NB 8.7.16).

There is a case for holding that independence of the world attained through the view *sub specie aeternitatis* involves one form of selflessness. This point is well made by Tilghman in his discussion of Wittgenstein's views. He says '...one becomes independent of... (the world)...by accepting the world as it is...' (WEA 74). 'The self represents the world as seen from a particular point of view' (WEA 60). Therefore the view *sub specie aeternitatis* involves seeing without the intrusive influence of self. It is a seeing from the standpoint of eternity and not from one viewpoint in the world. However, the fact that it involves a form of selflessness does not mean we can equate it with all forms of self-renunciation found in religious life.

Is the idea of independence of the world necessarily a self-renouncing *religious* idea? There is no reason to deny that it is so long as we recognise that it involves a very different sense of self-renunciation from that associated with many Christian traditions. This will become clearer after we have considered the case-study in Chapter 2. For the moment we can note that Monk talks of Wittgenstein's attempt to distance himself from his surroundings, including his fellow men. He quotes from a passage where Wittgenstein writes: 'Don't be dependent on the external world and then you have no fear of what happens in it...It is x times easier to be independent of things than to be independent of people. But one must be capable of that as well' (LWDG 116–7). This is evidence that independence of the world in Wittgenstein is a form of stoical *self*-concern with selflessness that is distinct from the identification with others that characterises many forms of religious self-renunciation.

The Wittgensteinian position seems to involve an uncritical application of Wittgenstein's idea of a view *sub specie aeternitatis* within a descriptive account of religious belief. There is a danger that this distorts many forms of self-renouncing faith by coming too close to conflating it with independence of the world – a point that will become clearer as this book unfolds. For the moment I shall now seek to give some examples of the way the idea of independence of the world is linked to self-renunciation in the Wittgensteinian position.

D.Z. Phillips' use of the perspective *sub specie aeternitatis* is imbued with reflections of Wittgenstein's idea of independence of the world. Phillips talks of 'seeing grace in all things' (RST 106) which he qualifies as an '...acceptance of the whole in which the self withdraws and all things are seen as a gift' (RST 104). This enables a way of looking

> *...at people and things in a way which includes the light and the dark...to see them with the whole of existence as their background; to see them sub specie aeternitatis* (RST 126).

Such a perspective provides, so Phillips claims, a *framework* within which the believer meets fortune, misfortune and the evil in life. It frames and determines the way the believer sees the world so that the saint and the atheist can be said to see different worlds. Such a 'seeing' involves no desire for an explanation in the face of pointless evil. It is said that '...this is what Simone Weil means by love of the beauty of the world'. This love encompasses both the beauty *and* the ugliness. The attitude of one who so loves

> *...is not determined by looking at things from the midst of them... (his/her)...attitude is other than the world's way of looking at things. His world is a different world from that of the man who sees objects from the midst of them'* (FPE 54–55).

The implication here is that the believer is freed from the constraints of a world in opposition to his or her aspirations. There is a release from dependence on everything turning out for the better.

It is important that we recall the context of this independence and how it fits into Phillips' wider account. Phillips thinks that a religion which tries to explain suffering is self-orientated and unauthentic. In contrast, he claims authentic faith is self-renouncing and involves living the acknowledgement that compensations for pointless suffering will never come. It involves accepting that death will not provide them. It involves accepting that'...in the ego there is nothing whatever, no psychological element, which external circumstances could not do away with' (CP 70). In his book *R.S. Thomas: Poet of the Hidden God*, Phillips claims self-renunciation involves '...the acceptance of all things as the will of God' (RST 87). It involves not

> *...seeking imaginary extra presences (but)...embracing the absences our questions come up against... (accepting)...the radical pointlessness in things* (RST 82).

Self-renouncing faith does not go on '...hoping for a reciprocating touch' (RST 152) nor does it '...desire a quasi-present God, a God whose existence can be felt if not proved' (RST 153).

In short, what the view of the whole, the view *sub specie aeternitatis* is supposed to enable is a sense of release from the cravings for the self's continuance. It provides the means for the self to come to terms with its eventual dissolution into nothingness.

We find the same sort of modelling of self-renunciation on the idea of independence of the world in Dilman. He says that if the claims of the flesh take

> *...root in (a person's) soul he can no longer look at things and respond to them from the perspective of the love of God* (SILAR 114).

Dilman portrays such a perspective as self-renouncing because it necessarily involves a diminishment of the sense of self and its replacement with a 'void' (SILAR 115). When there is this *void*, that is, this absence of the intrusive influence of the self, the world can be seen from the perspective of the love of God. Seeing from such a perspective is really a mode of seeing that is virtuous because it is unaffected by the self's natural orientation to the life of worldly pleasures. Action is no longer orientated towards the self's satisfaction and fulfilment. The person is now able '...to accept and love everything that happens' (SILAR 115) and by implication is independent of the world.

Such independence comes out clearly in his discussion of Socrates' view of the indestructibility of the soul. Dilman says this indestructibility can be understood as the possibility of a person finding eternal life. Such eternal life means being able '...to see things under the aspect of eternity, or *sub specie aeternitatis*' (SILAR 37). In turn, this involves detachment from time.

> *It is, of course, not easy to say what detachment from time amounts to, and what sort of temporal considerations a person becomes indifferent to...Certainly the desire for revenge, for consolation, for reward...are examples of what such a person becomes indifferent to* (SILAR 37).

These examples are all, I suggest, examples of an independence from the vicissitudes of this world and the need for a consoling explanation of them. Such independence entails that a person is indifferent to his or her own future (SILAR 116). The self arrives at an invulnerable sense of significance despite the miseries of the world: 'Nothing that happens within the world...can destroy that significance; it is invulnerable to the course of events in time' (PPL 127).

Another example that Dilman gives is taken from Eugene O'Neill's play *Long Day's Journey into Night*. The character Edmund has a glimpse of what comes into view when a person detaches himself from all those relationships that are subject to the incursions of the world and becomes independent of them. Edmund recalls how he was on lookout in the crow's nest of a sailing ship.

> *Then the ecstatic moment of freedom came. The peace, the end of the quest, the last harbour, the joy of belonging to a fulfilment beyond men's lousy, pitiful, greedy fears and hopes and dreams!...Like the saint's vision of beatitude. Like the veil of things as they are seem drawn back by an unseen hand* (PPL 126).

On another occasion

> *...I lost myself – actually lost my life. I was set free! I dissolved in the sea, became white sails and flying spray, became beauty and rhythm...I belonged, without past and future, within a peace and unity and a wild joy, within something greater than my own life, or the life of Man, to Life itself! To God , if you want to put it that way* (PPL 125–126).

Dilman agrees that this is what religious belief ultimately is. Man turns away from a daily life of 'survival of the self' (PPL 126). Man attains a perspective characterised by '...a sense of belonging timelessly ("without past or future") to something greater than his own life, and forming part of "peace and unity and a wild joy"...'(PPL 126). 'Only the beauty of each thing as it stands in the world as a whole matters; each thing takes its significance from that. Nothing that happens in the world, therefore, can destroy that significance: it is invulnerable to the course of events "in time", absolutely secure' (PPL 127).

To summarise, the term 'Perspective-Element' refers to the conjoinment of the 'perspective' with the idea of being independent of the world. Through seeing the *whole* of existence, a view *sub specie aeternitatis* in which the self is 'absent' and things and events are no longer judged in terms of what they contribute to the self's sense of security and wellbeing, pointless evil can be endured because the self is independent of the world.

The Anti-Metaphysical Element

The Wittgensteinian position stresses that the self-renouncing believer accepts the world and the contingencies within it as they are. He or she does not seek to explain or relate occurrences in terms of another, a metaphysical realm. In what follows I shall intend the term 'metaphysical' to means what is captured in the following quotation from Dilman:

> *... (I)f one turns the antithesis between what belongs to this world and what is supernatural into a contrast between what is here and what lies elsewhere, beneath, behind or beyond space, or into a contrast between what is now and what is to come later...then one will have made it into something with which it is impossible to have any relation that is not worldly* (SILAR 116).

Whether there is any intelligible sense in which these terms can have a purchase is not relevant to our immediate interests. Nor is the issue whether logically there is more than one contrast being made in this latter quotation. In what follows, sometimes 'transcendence' and 'supernatural' will have a meaning different from that of 'metaphysical'.

The Wittgensteinian position affirms that conceiving of God as a supernatural Being necessarily involves self-orientation. Fundamentally it is derivative of the desire for consolation of some sort.

This consolation is sometimes conceived of as a sense of security. Sutherland provides an interpretation of 'transcendence' which is distinct from the traditional metaphysical understanding, but one linked to the notion of self-renunciation. He insists, following Bonhoeffer, that transcendence is not to be understood as an '...attempt to have security from something visible' (GJB 117). For that would not be a genuine experience of God but a '...partial extension of our world' (GJB 119). Rather, transcendence is the '...freedom from self...the accuracy of vision which one has of another being' (GJB 122). He illustrates this with reference to Weil's idea of 'loving attention'.

The idea of security also underlies one central argument given by Dilman for the incompatibility of self-renunciation with the supernatural. Dilman points to the consolation involved in relating events to a supernatural realm by reference to self-orientated desire for a sense of a bond or contact to something else. If God is conceived of as somehow metaphysically present when we act, His presence would be 'something we can have in view when acting... (and)...this would change the character of our actions, they would no longer involve self-renunciation' (SILAR 115). If God is construed as present then

> *... (e)ven a martyr going to his death is supported by the bond he feels unites him to God. He does not feel alone and abandoned. This is not to deny that his action may not be completely selfless and disinterested. It may be so, but it does not diminish his sense of self. It does not replace it with a void. Christ felt no such support when he was nailed to the cross...he did not know that God was with him* (SILAR 115).

Something akin to this is found in Phillips' characterisation of authentic faith as not a seeking for a God '...whose existence can be felt...' (RST 153) or a 'reciprocating touch' (RST 152). As was noted in the discussion of the Anti-Consolation Element, there are problems with construing self-renouncing faith as necessarily involving no sense of a bond to anything at all.

Phillips has an argument which holds that the idea of a metaphysical God issues from both a desire for security and an urge towards self-centredness. He says that to see God as metaphysical is to see Him as a controller and therefore to want to see events bestowed with reference to oneself as the reason for their occurrence. 'It is precisely because there is no reason why things should go as they do in life that there is the possibility of seeing all things as acts of grace, as things bestowed without reference to oneself as the reason for their occurrence' (RST 82). Phillips thinks that there is something necessarily self-renouncing about 'accepting the radical pointlessness in things' (RST 82).

We can question whether it is a necessary condition of not seeing self as the reason for the occurrence of things that one finds everything pointless. As an example let us consider Diogenes Allen's theodicy as presented in his paper 'Natural Evil and the Love of God'. He argues that suffering can lead us to transcend our egotism.

> *(E)gotism is common. 'Why did this happen to me ? What did I ever do wrong ?' This is often said or felt with a sense of indignation, of outrage, of offence or self-pity... (But)...reflection can lead us to recognize more fully what we already know: we are material, and as a piece of matter we are vulnerable to injury, illness, and decay. To realize this is to realize our status, our place... (W)hen the flow of our self-regard is painfully interrupted, reflection may lead to a new awareness of our limitations, and it may lead to an act of acceptance of such limitations* (eds. M.M. and R.M. Adams, *The Problem of Evil*, Oxford: Oxford University Press, 1990, 193).

Allen rejects the position adopted by many philosophers, such as Hume, that regard evil as counter-evidence. Such a position precludes our seeing the religious lesson derived from suffering and merely reinforces our egocentricity. Now all this is something that Phillips is unlikely to quibble with. However, the significant point is that Allen does not couple this link between suffering and a diminution of our self-regard with a further link to the recognition of pointlessness. And I think this is a reasonably tenable position because there simply is no such necessary linkage. Coming to see that we are not at the centre of things is logically unconnected with any seeing of the nature of things as inherently pointless. Phillips is wrong to imply their conjoinment.

The issue as to what it means to remove a sense of pointlessness in the face of suffering is not unambiguous. Phillips does not really touch on this, but it is a genuine concern. Some philosophers have assumed that here 'not being pointless' requires a wholesale account of how a God of love can have sufficient reasons for permitting evils. However, this is not the only sense in which there is a justification in the sense of alleviating pointlessness in the face of evil.

Another sense of justification does not seek any wholesale reconciliation between propositions pertaining to attributes of an 'idealised' deity and those relating to the facts of evil in the world. Rather, it seeks merely to assist in the practical task of confronting evil by seeking to impart a sense in which a person can view suffering. One example by a philosopher is to be found in Robert Merrihew Adams' paper 'Existence, Self-Interest and the Problem of Evil' (in his *The Virtue of Faith*, Oxford: Oxford University Press, 1987). Broadly, his argument is that what we are attached to ourselves '...in a reasonable self-concern, is not just bare metaphysical identity, but also projects, friendships and...features of our personal history and character'. Many of the evils of the past have contributed to this sense of self-identity. He argues that Helen Keller would have lived a very different life had she not been born blind. When evils are viewed in relation to the goodness of life as a whole we cannot, out of reasonable self-interest, blame God.

Phillips could criticise Adams for the role self-interest plays in his argument. Against this, it could be claimed that the form of self-interest Adams highlights is very different from gluttonous self-orientation. The example discussed earlier from Diogenes Allen represents a distinct development in this form of justification. It seeks to express and advocate a specifically Christian appreciation of suffering. (Adams, in contrast, appeals to considerations which a lukewarm-warm even non-religious person might recognise.)

A further aspect of the anti-metaphysical strain in the Wittgensteinian position is to be found in assumptions pertaining to the concept of the 'power of God'.

One line of thought that illustrates this is Rush Rhees's dismissal of the attempt to understand the miraculous in terms of quasi-causal power. For this is in reality an attempt to utilise a metaphysically conceived God as a sort of mechanism to bring things about. Rhees says:

> *...a quantitative comparison between the physical effects of God's power and the physical effects of anything else would be a pretty unholy sort of thing...* (WA 113).

One aspect of this unholiness can be grasped by reference to what A.P. Shooman, a former student of R.W. Beardsmore, says about a causal explanation of miracles:

> *...To ask for a causal explanation of a miracle is to put oneself beyond the point where talk of the miraculous can mean anything; for asking what Christ does involves a failure to recognize who Christ is. The assumption that lies tacitly rooted in this way of talking is that the difference between, for example, myself and Jesus Christ is a matter of degree. Whereas I cannot raise the dead, He can. It becomes a question of who can do what. There is an unholiness in this way of talking* (MRB 46).

The unholiness consists in elevating the self through clinging to the illusion that it is only a degree or so removed from God. The assumption here is that the logical status of God's power is merely a consequence of its physical effects in the world. But why should we think this?

Let us look at D.Z. Phillips' use of this power-argument in his criticism of Geach (DI 21 ff.). He claims that to associate God with 'power' is to merely offer a worldly incentive to a person as to why he or she should believe in God. Such incentives can only, according to Phillips, result in a policy for self-security, 'a means to get what one wants in this dangerous world' (DI 29) and not a self-renouncing faith in God. The problem with Phillips' claim is its uncritical assumption that something is a worldly power merely because it competes with other worldly powers or aims at a victory which triumphs over other powers in this world. In his paper 'The Perfect and the Particular' (Inaugural Lecture, King's College London, June 1994), Paul Helm criticizes the assumption that something is a worldly victory by virtue of what it competes with or aspires to triumph over. He points out that what determines whether something is a worldly victory is the nature of that victory itself.

Phillips seems to be led astray by assuming that things only logically compete when they are in the same logical category. Yet this is a questionable assumption which is criticized in Charles Taylor's essay 'Rationality' (in eds. M. Hollis and S. Lukes, *Rationality*, Oxford: Blackwell, 1982). Taylor shows how concepts belonging to another, incommensurable belief-system, which he assumes is the sort of contrast posited in Winch's famous essay 'Understanding a Primitive Society', can yet pose a logical threat so that we can intelligibly talk of 'valid transcultural judgements of superiority'. The difficulty in adopting Taylor's illustration here is that it is based ultimately on how '...theoretical cultures score successes which command the attention of atheoretical cultures' (104). In other words, Taylor only finally manages to show how worldly technological powers triumph over incommensurable non-technological ones. In contrast, what we are seeking is some purchase in the idea that non-worldly powers, though incommensurable with worldly ones, can yet be held to triumph over them.

Such an illustration is found in Weil:

> *The things of this world exist. Therefore I do not detach from them those of my faculties that are related to existence. But since the things of this world contain no good, I simply detach from them the faculty which is related to the good, that is to say, the faculty of love* (quoted SW 200).

The contrast here is between the 'things of this world' and 'the good'. The 'things of this world' are in a different conceptual space from 'the good', as is shown by the sense in which the 'things of this world' can be said to have a relation to 'those of my faculties that are related to existence'. 'The good' has no such relation. Yet the concept of 'the good' is logically something which can affect assessment of the things of this world.

Thus the use of the idea of 'power' in connection with God, even when that is used in a way in which it is logically thought to compete with worldly power is not sufficient to make God logically into something comparable to a worldly power. This will be illustrated further in Chapter 2.

A further example of the Anti-Metaphysical Element in the Wittgensteinian position concerns its stress that self-renouncing religion can have no place for the idea of a continuation of the self beyond death. Wittgenstein himself clearly found no place for the idea of a continuation of the self beyond death. In the Tractatus he says that

> *...this assumption completely fails to accomplish the purpose for which it has always been intended. Or is some riddle solved by my surviving for ever? Is not this eternal life as much a riddle as our present life?* (TLP 6.4312).

Wittgenstein later attacked the intelligibility of the idea of surviving death as it is often conventionally expressed.

> *Philosophers who say:'after death a timeless state will begin'...and not notice that they have used the words 'after' and 'at' and 'begins' in a temporal sense, and that temporality is embedded in their grammar* (CV 22).

Many of the conceptual difficulties associated with the idea of a continuation of the self beyond death, including the intelligibility of the notion of disembodied existence, are dealt with in D.Z. Phillips' *Death and Immortality*. Phillips' distinctive contribution is to be found in the way that he, perhaps more than any other exponent of the Wittgensteinian position, has pointed out not merely conceptual problems but also objected to the self-orientated dimension supposedly inherent in the idea.

The essential point in Phillips' descriptive account of the concept of immortality is that it does not imply an extension of life. Immortality is not '...more life but this life seen under certain moral and religious modes of thought' (DI 49). It refers to a person's '...relation to the self-effacement and love of others involved in dying to the self' (DI 54). Dying to the self is what authentic faith is about (CP 70; RST 102).

What is the connection, in Phillips' thought, between authentic religious belief as involving dying to the self and the idea that death is the cessation of the self? The recognition of death as a termination of the self is said to be self-renouncing because it involves acceptance of the truth that there are never to be compensations for misfortunes suffered in this life. But if we think of surviving death, Phillips says '...the lesson religious believers see in death is lost, since death no longer reveals the fact that there is to be no compensation, but is seen as an additional fact for which compensation is sought' (DI 53).

It might be questioned why a cessationist view should be assumed to be automatically non-consolatory. In a passage from Derek Parfit we find, apparently, a different view:

> *After my death there will be no one living who will be me...Though there will later be many experiences, none of these experiences will be connected to my present experiences by chains of such direct connections as those involved in experience-memory, or in carrying out of an earlier intention* (*Reasons and Persons*, Oxford: Clarendon Press, 1984, 281).

Parfit stresses that he finds his view, a view which, superficially, does accept the termination of the self at death, actually '...liberating and consoling'.

Phillips could no doubt reply to this apparent counter-example. He might point out that it is really a case of someone who, faced with the self's termination, yet manages to find some solace in thinking that something of the self – although that something is much less that direct continuity of consciousness – will actually persist after physical death.

Although this stress on accepting the finality of death is true to some religious traditions, we shall see in the next chapter that it is far from universal.

None of what has been said above should give the impression that I am affirming that *all* self-renouncing conceptions involving the concept of transcendence involve a metaphysical conception of God. Rather, we must recognise that there are some which are distinctly not metaphysical. One example is to be found in Dilman's distinction between 'religious' and 'philosophical' transcendence as explicated in some parts of his paper 'Metaphysical and Religious Transcendence' (SILAR 109–115). Dilman says that when the Christian God is described as 'beyond the senses' it is the world of the senses that is in question. In Christianity the world of the senses does not mean

> *...those things we may be said to perceive with the senses (but)...the world in which we seek the satisfaction of sensual pleasures, bodily appetites...the desire for riches, power and fame...the world where concern for these pleasures overshadows all other concerns* (113).

I think that there are cases in which Dilman is right to say that turning to God, in terms of this contrast, is a matter of renouncing all these amenities. Another example of a non-metaphysical sense of transcendence is elucidated by David Cockburn ('The Supernatural', *Religious Studies* 28, 1992) and also by Peter

Winch in his *Simone Weil: The Just Balance*. They show us a sense in which the language-game of the supernatural is based on reactions of wonder and awe. Such reactions involve recognition of a particular kind of virtue – they do not point to another realm behind, above or in addition to this one. In Weil's terms, a 'supernatural virtue' involves a refrain from using powers that the self could, in worldly terms, legitimately utilize to secure its desired outcomes. Cockburn's examples include a man going to considerable length and great risk to help a stranger. Winch says of such cases:

> *Where...the agent expects some advantage from an action...we can point to the source of energy which makes possible the action...in other cases, we cannot* (SW 208).

In these latter cases, where the self seeks no advantage, the source of the energy is supernatural, 'outside the world'. Cockburn says that here the thought is '...that this is impossible – that no explanation in earthly terms could show how such behaviour can occur' (290). We cannot account for why an agent did such-and-such a thing in terms of advantage or satisfaction. Here the role of the 'supernatural' (or the transcendent) is not to provide a justification but arises from a sense of wonder at cases where no such explanation is to be had.

It is worth emphasising that recognition of cases such as provided by Dilman, Cockburn and Winch in the last paragraph does not commit us to any views about how general they are in religious life. It is part of the purpose of this book to resist the temptation to find here a necessary condition of self-renouncing faith.

Conclusion

In this chapter I have provided an outline of five interrelated elements which are widely found in Wittgensteinian writings on religious belief. These elements are: the Absoluteness-Element; the Anti-Consolation Element; the Unreflectiveness-Element; the Perspective-Element and the Anti-Metaphysical Element. Taken together these elements comprise what I have termed 'the Wittgensteinian position'. I do not claim that each of the philosophers I have cited at various points in this chapter necessarily subscribes explicitly to each of the above elements. Rather, when most or all of these elements combine then we find a presupposed model of just what can count as authentic, religion. Such a view of authenticity is implicit in many philosophers inclined to a Wittgensteinian approach.

The presupposed model hangs together in roughly the following way. The self-renouncing believer's faith is absolute; it has no further ends and is not held for the acquisition of any consolatory benefits. At its most genuinely self-renouncing, such belief excludes reflective engagement with one's values. The self is immersed in an unreflective perspective on life seen *sub speciae aeternitatis* and thereby imbued with a sense of being independent of the world. In this there is no orientation to the metaphysical dimensions traditionally held to be something integral to religion. Such an orientation is presumed to cancel out the self-renouncing character of authentic religion.

There are influences within the Wittgensteinian tradition that push us toward this model. The aim of this work is to critically examine it in the hope that a clearer awareness of the nature of these influences, these Wittgensteinian *values*, can contribute at least something to inspiring us to refine our practice of a descriptivist methodology and make it more open to the complexity of what is found within religious life.

2 Self-Renouncing Religious Belief in the Novels of Georges Bernanos

Introduction

The purpose of this chapter is to provide a contrasting conception of the nature of self-renouncing faith to that outlined in Chapter 1. This will involve an analysis of the novels of Georges Bernanos.

The French writer Georges Bernanos (1888–1948) ranks as a major exemplar of the so-called 'Catholic Novel'. From his first novel *Sous le Soleil de Satan* (1926) onwards he struggled to provide twentieth century readers with a presentation of the supernatural at work in the fabric of this world. In that first novel we find a tendency towards a faith that relies on miraculous occurrences and the drama of the priest Donissan's struggle with the powerful reality of evil.

In his subsequent writings we find a shift away from a dramatic, in some ways stereotypical, view of self-renunciation as simply a battle against temptation. In *La Joie* (1929) we find an emphasis on the childlike simplicity of the central character, Chantal, coupled with an emotional fullness that serves as a contrast with the inner emptiness of the other characters, consumed to various degrees by evil. Despite the progress in this novel, Bernanos yet came to feel that Chantal's character was presented as too passive. We can certainly concur because in her case self-renunciation is presented on the model of a receptacle given a quasi-mechanical infusion of grace.

It is with the novel *Journal d'un cure de campagne* (1936) that we find a distinctive model of self-renouncing faith centred on the idea of *L'esprit de l'enfance*, the spirit of childhood [1]. This work is expressive of the form of sainthood we find in such Carmelite saints as St Therese of Lisieux. Bernanos came to recognise the contrast between her sanctity and that of the more dramatic, individualistic and assertive type which his first novel had concentrated on He spoke of the '...supernatural youth incarnated by... St Therese, to the scandal and trial of every set of fanatics... who would like to turn the Church into an austere cemetery, instead of the garden in bloom' (from an interview quoted in Robert Speaight, *Georges Bernanos*, London: Collins & Harvill Press, 1973, 266).

In this chapter I shall attempt to show something of the nature of Bernanos' conception of self-renouncing faith. Most of my account will be based on the *Journal d'un cure de campagne*, though some points will be supplemented by reference to other works, such as the play *Dialogues des Carmelites* (1949). To understand the distinctiveness of Bernanos' view of self-renouncing faith we have to understand the idea of the spirit of childhood. A fuller account of this concept will be presented in the following sections.

Bernanos and the Unreflectiveness-Element

It is appropriate to start exploring Bernanos' view of self-renunciation with reference to the Unreflectiveness-Element because the novel *Journal d'un cure de campagne* is a work in which the life of the saintly Priest of Ambricourt is portrayed as saturated with reflection about himself, the world, the obstacles he faces, and the unattainability of a sense of God. The very fact that the novel is written in the form of a diary emphasises this element of reflectiveness. The Unreflectiveness-Element is, as we shall see, incompatible with Bernanos' conception of self-renouncing faith.

It will be recalled from Chapter 1 that the Unreflectiveness-Element implies that reflectiveness and articulation involve consolation and are therefore avenues for the self to intrude. For self-renouncing faith to be free of the intrusive influence of self, there must be an absence of reflectiveness: action must be unreflective, immediate and free of all intervening articulation. In Bernanos we find three points which are at variance with the Unreflectiveness-Element: (a) That self-renouncing faith is not an absence of self but a matter of the nature of the self that is present. This is a self characterised by a form of childlike innocence and humility devoid of self-assurance; (b) In this form of childlike innocence there is no immediacy nor the lack of any intervening process between discernment of a situation and self-renouncing response; (c) reflection is not incompatible with such an innocence. There is no unquestionability; the Priest of Ambricourt can doubt his mission and the existence and purposes of God without compromising his innocence.

We shall start with the first point. In Bernanos self-renouncing faith is not an absence of self but a matter of the nature of the self that is present. The Priest of Ambricourt is characterised by a form of childlike innocence. This childlike innocence is not at all comparable to innocence as necessarily devoid of uncharitable motives or unacquainted with misdeeds and regret. Rather, the nature of Bernanos's view of innocence is centred on a humility devoid of worldly self-assurance. Self-assurance involves a sense in which the self, enmeshed in its own capacity to command or captivate others, also remains at a distance from others' ultimate needs.

The Priest of Ambricourt possesses the spirit of childhood. Bernanos equates the spirit of childhood with what he calls 'the spirit of poverty' (J 237). What is this 'spirit of poverty'? Essentially, for Bernanos, a self that has this spirit of poverty is one characterised by a freedom from valuing things and relationships from a standpoint of what they bring in terms of personal self-esteem and satisfaction.

Bernanos builds his portrait of the Priest of Ambricourt by alluding in some detail to his social background. The point of this is not to outline what genealogy a saint must have. Rather, the poverty of the priest's origins is indicative of how he 'stands' in terms of relationships with other. It conveys the character of his orientation to the world.

The Priest of Ambricourt is not only of peasant stock, as is also his elder colleague from a neighbouring parish, the genuine and street-wise Priest of Torcy. He is also of the humblest level of the French peasantry. Whereas Torcy's family were successful farmers in the prosperous Flanders countryside, the Priest of Ambricourt comes from a class of landless alcoholics and spendthrifts with little in the way of dignity or moral character to maintain. When his father died and his mother became ill, he lived with an aunt who kept a pub where they sold gin to miners who were too poor to go anywhere else. The child would crouch, terrified, behind the bar on the few rotting floor boards. In such circumstances he came to know an acute form of distress.

> *Amongst the poor...distress is not shared, each creature is alone in his distress; it belongs only to him like his face and his hands* (J 48).

The Priest of Ambricourt has known this state of distress that has ceased to reason and that is unacquainted with hope.

This background has given the Priest of Ambricourt a sensitivity which is not available to those who have the self-assurance which the privilege of money bestows. This priest, coming from the lowest social stratum, has never had anything to defend in the concourse of human hierarchies. In not being raised in the tradition of possession he bears no grudge against the world that fails to deliver the privileges and status that so many adults consider tempting. He has never known bitterness such as that of the land owning peasant against the land itself for eating up his strength (J 33). For Bernanos the spirit of childhood is akin to the spirit of poverty in that it makes possible a form of relating between the self and others which is devoid of dominance, prudence and the sense of expectation that, if thwarted, gives rise to bitterness.

Dominance, for Bernanos, is really only a matter of utility, of one self using another for its consolation or sense of self-esteem. A lack of dominance makes possible a form of humanity that achieves a sense of the mystery and the depth of our ties with each other. It makes possible a form of compassion that one who is tied to his or her stake in worldly hierarchies of influences and power cannot achieve. The spirit of poverty that Bernanos is seeking to illustrate, is, for him, something that makes possible a heightened form of fidelity to, and solidarity with, others. In his diary the priest writes: 'Therein lies my whole strength, the strength of children and weaklings' (J 53).

Prudence is seen by Bernanos as a further obstacle to the innocence that is part of the spirit of poverty and of childhood. The Priest of Ambricourt is criticised by church leaders for not being more ready to accommodate himself to commercial dishonesty among some parishioners. Despite ill-health, he adheres without compromise to the hard schedule of visits he has drawn up to ensure regular contact with members of his far flung country flock. His colleagues consider this an excessive duty. He spends endless hours on the road. Often the stomach pains become acute when he is on his bike. But the spirit of poverty does not count the cost of commitment. His elder spiritual guide, the priest of Torcy, tells him that he pushes himself too far (J 79). But Torcy also recognises that giving no heed to the self's security is part of the priest of Ambricourt's innocence. Torcy is Bernanos' mouthpiece when he says: 'There's a kind of supernatural laziness which comes with age, disappointment, experience…Old priests are as hard as nails! Prudence is the final imprudence when by slow degrees it prepares the mind to do without God' (J 79).

To summarise, what we find in Bernanos is not a conception of self-renunciation as an absence of the intrusive influence of self. Rather, it involves a particular character of self, one imbued with a particular form of childlike innocence. This spirit of childhood is found in the spirit of poverty, a spirit that is part of a self that is not tied to worldly standards and aspirations of power, dominance, self-esteem and prudence. The self that is enmeshed in these is a self that, despite apparent comradeship and the display of proprieties in social intercourse, is really at a distance from others. The absence of any thought of self-enhancement, for any prestigious 'place' for the self in the world, which characterises the spirit of childhood is something Bernanos sees as a pre-requisite to being sensitive to the fragility and the needs of others.

Turning next to the second point delineated at the beginning of this section, we need to be clear as to how Bernanos' conception of the spirit of childhood is at variance with the Unreflectiveness-Element's stress on the unreflective immediacy and lack of any intervening process in self-renouncing religious and moral responses. The Priest of Ambricourt does not have the practical and capable fluency of Winch's Good Samaritan. The latter grasps the situation in an instant and responds. In contrast, the priest is often uncertain as to how he should act or whether his devotion to God is of the right nature. He is hesitant, frail and uncertain. He struggles to try to find a sense of vocation and sees himself 'simply as an instrument used by God'. He is physically weak, suffering from undiagnosed cancer, and socially inferior. His sense of inadequacy is accentuated by failures in his parochial duties. He embarrasses himself and others by sometimes prolonging his visits and never knowing how to leave gracefully.

In recounting a conference of the clergy he recalls how mystified he was by the fluency with which others could use words and command attention (J 34). Generally he finds he cannot string together ten words without stumbling (J 125). He recounts how he gives extra money of his own to a woman engaged by the local Comte to give weekly domestic help to him as parish priest. Her peasant ingratitude is something he feels he must overlook because he blames himself for the awkwardness in the manner in which he makes his payment.

But this simplicity, this muddled uncertainty about whether his own acts and attitudes are appropriate, means the Priest of Ambricourt can discern the good in others.

One startling example of this uncertainty is found in his reaction to the diagnosis that his illness is terminal cancer. At that moment the doctor was watching him, waiting for his reaction.

> *I should have gone, but I couldn't...I waited for the gift of one word, one word that a priest might utter, for that one word I would have given my life...* (J 234).

The words that would project an image of practical priestly dignity and piety elude him and he leaves in tears. Bernanos himself wrote:

> *Saints are not heroes like those of Plutarch. Heroes give the impression of being supermen, but the saint is not a super-man. He embraces his humanity and tries to live it completely...following Him who was most perfectly human* (*La Liberte Pour Quoi faire?*, Paris: Gallimard, 1953:286).

The Priest of Ambricourt's impracticality can be contrasted with that of his alert, street-wise elder colleague, the Priest of Torcy. Torcy knows how to deal with any practical situation. But during the course of the novel he comes to discern how his capable practicality and strength tends to edge too closely to pride rather than sanctity.

> *No, he (Torcy) couldn't see me. The mere intention to convince me would not have set such tragedy in his eyes. He was battling with himself, against another self, crushed a hundred times, vanquished a hundred times, still in rebellion...* (J 51).

This is Torcy's recognition of the distance separating himself from that ideal spirit of poverty and capacity for pity. Torcy says: 'Pity is powerful and devouring. I don't know why we always think of it as something rather snivelling and silly' (J 52). Bernanos saw in the assured and proud capability of Torcy, a man who knows immediately what response is required in any given situation, something that falls short of the of the self-renunciation exemplified in the saintly Priest of Ambricourt.

The third point highlighted at the beginning of this section as contrasting with the Unreflectiveness-Element was the place Bernanos allows for reflectivness in self-renouncing faith. The Priest of Ambricourt reflects upon his own motives, actions, failures and sense of guilt. He also reflects on whether he has a mission and whether there is a God who inspires it – his faith in God is, for much of the novel, far from unquestionable. During one intense period of physical and mental suffering he even comes to doubt whether there is a God that knows of his suffering (J 126).

The Priest reflects on his own actions, motives and failures. The central example in this connection is the diary which he keeps. 'This diary is of immense help in forcing me to see my own share of responsibility in so much bitter disappointment…' (J 125).

It might be objected that the diary format is merely the device used to structure the novel. In other words, that the novel 'works' better as a first person narration in this format. We cannot, it might be urged, deduce anything about the nature of the spirituality concerned by reference to it. Against this objection I would urge that reflectiveness is enshrined in the form of religion exemplified by the Priest of Ambricourt. It is not something solely to be attributed to the diary-format. In support of this it is worth recalling that many Catholic saints have kept journals. A classic example is that of St Teresa of Lisaux. Bernanos is known to have been impressed by her. It is likely that the Priest of Ambricourt's self-renouncing faith has been partly modelled on that of this saint.

As well as reflecting on his actions, motives and failures, the Priest of Ambricourt also reflects on the *purpose* of religious life. He articulates things in order to try to find a point in it all. One example of this is his reflection on the value of developing an 'inner life'. He believes that most people live shallow lives.

> *...Therefore when death has bereft them of all artificial props with which society provides such people, they will find themselves as they really are...horrible, underdeveloped monsters, the stumps of men* (J 94).

The lack of such an inner life is connected with lack of love. The Priest of Ambricourt makes the following claim about those damned to hell, those who have not developed an inner life. 'Hell is not to love any more' (J 140). Hell is the state in which the individual has nothing to share with others.

Elsewhere the Priest contrasts the inner life with the sort of life that is lived solely in terms of 'social discipline and habit' (J 95). The Priest sees this most clearly exemplified in the professional attitude of the typical French lawyer. Such a life can be '...expressed in very few words' (J 95). I think he means by this that such a life is lived in maintaining and reinforcing a sense of the dignity of the self in such a way that dignity becomes a barrier to appreciating the reality of other people. To this sort of professional approach, a person in distress is reduced to being an opportunity for the professional self to consolidate its prestige.

Bernanos is also implying that such professionalised practicality and dignity is suspicious of reflectiveness. It resists all attempts at analysis which might show that its practitioner is ultimately not unrelated to the unfortunate individuals it confronts.

The point to note in this is the fact that the Priest of Ambricourt sees in this a *justification* for living the religious life. The Unreflectiveness-Element implies that such a justification is self-affirming because it is uncritically assumed that it involves the self assessing what benefits the religious life provides. But here the Unreflectiveness-Element is shown to be over simplistic.

In its defence it might be claimed that the justification at issue here would only satisfy someone who already is a certain sort of Christian believer. In reply, the fact that a justification is needed suggests that belief here is not a wholly unreflective reaction.

The Priest of Ambricourt's reflection on the idea of an inner life does not consolidate his self. Rather, it is the vehicle for expressing the selfless character of the self he is seeking to be (and, for Bernanos, the self he already is). Through this process of reflection he comes to see more clearly a way of being involved with others that does not confine, distort or ignore the reality of their needs for the sake of enlarging one's own self-esteem. The religious life is the means to that form of 'being' in relation to others.

Two final points need to be noted from this latter discussion on the idea of an 'inner life' as found in the novel.

Firstly, one implication of the Unreflectiveness-Element, namely that we can straightforwardly segregate belief-*in* from belief-*about*, is untenable. In the case of the Priest of Ambricourt's reflection on the 'inner life' what we find is a case of belief-about which is integrally part of belief-in, that is, belief as *lived* in religious life.

Secondly, the above example enables us to question whether all religious 'pictures', as adhered to by the self-renouncing believer, are wholly pictorial and unreflective. The idea of an inner life, and progress in attaining it, does have a regulative dimension. It is something the Priest of Amricourt returns to a number of times. It provides a means for him to articulate what is, as it were, happening when one seeks to live the religious life. It enables him to conceptualise things in terms of a self-development leading to an ever more intense exemplification of the 'spirit of childhood'. But this 'picture' of an inner life does not achieve this regulative character by being wholly pictorial. It is something that is necessarily encased in words. It has its context in religious life in a reflective activity involving verbal articulation.

Bernanos and the Perspective-Element

The Wittgensteinian Perspective-Element outlined in Chapter 1 concerns the unreflective view of the whole of life in which the self attains an *independence* from the world, a stance unaffected by the onslaughts of suffering heaped upon it by the course of events in life.

In this chapter it will be argued that the Bernanosian view of suffering is not a form of 'independence from the world' and is incompatible with the Perspective-Element. The latter is not seen as either intrinsic to self-renouncing faith nor inherently self-renouncing.

We begin with what Bernanos sees as the ultimate cause of suffering. Hebblethwaite rightly claims that:

> *(Bernanos') whole work is directed to the conquest of despair, and in his saints he points to a way of making sense of and overcoming angoisse...* (Quoted in P. Hebblethwaite, *Bernanos*, London: Bowes & Bowes, 1965, 116. Hereafter 'BH').

For Bernanos one of the principal sources of despair is the inability to accept suffering.

> *The disgrace of the modern world is not that it should suffer but that it should suffer in vain. All suffering is vain and intolerable for someone who lays the blame on the obstacle, like a child who beats with his fist an object on which he has hurt himself...* (BH 57).

Bernanos believed that the modern world saw suffering as intolerable because it saw it as a foreign imposition, an intrusion from an alien, external world that threatened the dominance of the *individualised* self.

According to Bernanos' view, the self-renouncing believer sees suffering as confronting not the individualised subject but the collectivity of the Church. The Priest of Ambricourt ultimately (though at many points tempted and deflected) refuses to view it as striking him as an individualised, isolated victim. Rather, he struggles to see his own suffering as an expression of, even a means toward, his extension of solidarity with others.

Solidarity is a central theme in Bernanos. For him the self comes to greater knowledge of itself through its ties with others. This is well illustrated in the *Dialogues de Carmelites* where the Prioress talks of the way we first see our own infirmities through others. Thus the knowledge that charity makes possible is essential to human solidarity. Bernanos does not see such solidarity merely as a matter of 'unlocking' the depths of the other. It is also a matter of perceiving oneself aright *through* the other. For example, the Prioress in the *Dialogues des Carmelites* says that 'it is first in others that we discover our own extreme sickness' (DC 39). She says that the person who, through a false charity, uncritically accepts the other's shortcomings 'does no more than break the mirror that he may not see himself in it' (DC 39). Charity towards others is a vehicle for accurate knowledge of the self. Without the solidarity with others that it makes possible, we are imprisoned in our narrow lives of half-truths, clichés and oversimplified ideas. Not only is our view of the world a confined one, but so is our understanding of ourselves. It is not through any penetration into an inner self that the diary enables the Priest of Ambricourt to come to terms with and understand himself; it is through its enabling him to be more fully perceptive in his encounters with others. In turn, those encounters lead him to a deeper understanding of himself and his mission. (This point can be illustrated further by referring to the discussion, in the section on Bernanos and the Anti-Consolation Element, of the self-compassion achieved by the Priest of Ambricourt through his dealings with the Comtesse and, in particular, his short friendship with Monsieur Olivier.)

Some commentators claim that Bernanos' *Journal d'un Cure de Campagne* illustrates the doctrine of Vicarious Substitution. Vicarious Substitution is defined by Richard Griffiths as '...the suffering of one of God's creatures to expiate the sins of the others' (*The Reactionary Revolution*, London: Constable, 1966, 201). The usual elucidation of this doctrine involves a metaphysical exchange. Thus in the English novel *Descent into Hell* by Charles Williams we find Peter Stanhope assuming the fear that bedevils the life of Pauline Ansruther so that she may experience the joy of liberation. Here the emotional state of fear is supposed to be *transferable* by some metaphysical means from one individual to another. Commentators such as Ernest Beaumont apply this type of account to Bernanos:

> *The* Dialogues des Carmelites *illustrates how Blanche's fear of death is assumed by the old prioress who dies, not her own death, but, through the fear that she exhibits, Blanche's. As Soeur Constance says: 'We do not die each for ourselves, but each for another, or some even in the place of others...'* ('Georges Bernanos', in *The Novelist as Philosopher*, Oxford: Oxford University Press, 1962, 39).

Beaumont claims that this form of metaphysical substitution also occurs in the *Journal d'un Cure de Campagne*. He points to what happens after the Priest of Ambricourt's confrontation with the Comtesse.

> *... (She) dies during the night that follows... (and)...is freed from the state of revolt in which the death of her infant son years before plunged her. The priest, though, has an inkling of the exchange that has taken place: 'The hope that was dying in my heart has reflowered in hers, the spirit of prayer which I thought I had irremediably lost God has given to her, who knows, perhaps in my name'.*

Beaumont's justification for claiming that Bernanos here conceives of the doctrine of Vicarious Substitution as a matter of a *metaphysical* exchange is the fact that emotional states are somehow transferred from one character to another. But do we *have* to think of this as some kind of psychic transfer?

There is another way of understanding the transferability in question. The fact that the Priest of Ambricourt assumes the emotional states of others is indicative of his deep feeling for others and their plight. Thus, after the Comtesse's death he 'spent the first hours of this horrible day in a state very like rebellion'. 'In...mirrors I saw a face which seemed disfigured less by sorrow than by fear...' (J 151). Bernanos uses this image of transferability to express not the idea that emotional states can, like photons of light, travel through space and literally move from one person to another. Rather, if we have a deep enough feeling for another person then certainly we will assume some of their emotions. For example, parents can weep and become fearful when their sick child is distressed. The Priest of Ambricourt is one who so closely identifies with others in this way that he comes to share their griefs, fears and pains through *pity*. The French word *pitie* means both pity and compassion (though in what follows I shall use only the English term 'pity').

The Priest of Torcy expresses the power of pity alongside the common man's instinct to draw away from it:

> *...Pity is like an animal...Pity is powerful and devouring... I don't know why we always think of it as something rather snivelling and silly. One of the strongest passions of men – that's what it is...* (J 52).

For most people, too great an immersion in pity threatens their sense of self. One of the great ironies of the novel is that the strong, healthy Priest of Torcy, with his great knowledge of worldly affairs and yet sincere aspiration for sanctity, can only with distinct unease approach the strong passion of *pity*, even though he recognises its roots in the life of Christ. Yet it is the weak, sickly, often inept Priest of Ambricourt who is able to handle it. In him, this outwardly unpromising, anaemic figure there grows the courage to overthrow the throes of a certain form of attachment to self and to live with this powerful, devouring thing that is pity, this thing that for Bernanos represents the very scourge of the worldly self.

The Priest of Ambricourt's capacity for genuine pity permeates the entire novel. He thinks of the old, atheist doctor with compassion: 'And I heard, or thought I heard, the groaning of so many men, their dry sobs, their sighs, the rattle of their grief, grief of our wretched humanity, pressed to earth, its fearsome murmurings' (J 146). He sees in the arrogant Comtesse who is wreaked by hatred of God, the pain and wretchedness of despair.

> *As I listened sadness overwhelmed me, indefinable sadness against which I felt quite powerless. That may have been the worst temptation in my life. But then God helped me. Suddenly I could feel a tear on my cheek, a single tear, as we see them on the faces of the dying, at the furthest limit of their griefs. She watched the tear fall* (J 141).

In the above we see that in his pity for her he is tempted not to challenge her, to leave her view of things untouched. But then he recovers himself and persists with the confrontation, the struggle to expose the insufficiency of the hate that ruled her heart. His pity is expressed in the tear that runs down his face. The Comtesse, after years of arid hatred, at last sees before her someone with a genuine capacity to pity, to compassionately understand her grief. The depth of his pity also comes out when he later visits the family and sees her dead body.

> *I even saw a little mark, a scratch which I had noticed yesterday, as she pressed the medallion to her heart. The thin strip of lint was still attached to it. I don't know why that broke my heart. The memory of her struggle before my eyes, that fight for eternal life from which she emerged exhausted...became painfully vivid, shattering...* (J 154).

Notice here the attention to details of the particularity of the other person.

There is also the pity the Priest of Ambricourt feels for Torcy, for the latter's realisation of his own enmeshment in pride.

> *He was as though tearing himself with his own hands...His eyes rested on me and I was ashamed of my little troubles...* (J 51–52).

The Priest of Ambricourt's capacity to pity and to identify with others extends to all those he comes into contact with. Even those, such as his mean and suspicious housekeeper, who he experiences a natural dislike for.

It is the knowledge of distress and the capacity to pity that enable us to appreciate the character of Bernanos' conception of solidarity with others.

We do not need to think in terms of a metaphysical transfer to appreciate this character. Consider the words of the Priest of Ambricourt, written in his diary:

> *True pain coming out of a man belongs primarily to God... I try and take it humbly to my heart...to make it mine, to love it...I really 'commune' with his pain* (J 72).

Here what we have is no psychic transfer of pain-states but an intense sense of being bound to others founded on an appreciation of distress and a capacity for pity.

At this a critic might ask: 'Just what good is pity? What difference can it make?' For Bernanos, it does make a difference. In the Journal we find an illustration which goes some way to showing us the nature of that difference. Consider the example discussed above of the Priest of Ambricourt's meeting with the Comtesse. In seeing the tear running down his face she discerns his pity. She no longer finds before her one who is there to challenge her self for dominance. Rather, she finds a self with the capacity to be receptive to her distress.

Why should mere receptivity of this kind make a difference? Here we touch on something that relates to what the Wittgensteinian notion of the 'natural history of mankind'. Human beings simply do find a strength, and a capacity to readjust their lives, in recognising they have the pity of others. The Priest of Ambricourt himself, when near to death, shudders in the face of that death and says: 'God might possibly wish my death as some form of example to others. But I would rather have their pity…' (J 249). Pity here is a comfort against distress. The Priest wishes it for the alleviation of misery that it provides: in his humility he is uncertain of his own capacity – unsupported by the pity of others – to enable him to meet his own death.

It must be stressed that the pity Bernanos is seeking to show us is different from that secular notion of pity as a form of condescension towards the sufferer. *That* form of pity is inherently linked to one self's aspiration to dominate and suppress another. The Bernanosian concept of pity involves communion with another's pain. Underlying the Priest of Ambricourt's capacity for pity is his awareness of the distress infecting so many individuals. He sees that outside the Church distress is not shared, but that each individual is left abandoned in his or her affliction (J 47).

Through pity, the solidarity fostered by the example of the saints who are imbued with the spirit of childhood, suffering should not be seen as striking at the separate individual, segregated into a discrete isolation. Suffering is a condition that afflicts the Church, the collective of all individual selves, bound together in Charity, nurtured by the example of those saints who appreciate the misery and distress of life without God.

To summarise things thus far, we have explored Bernanos' view of suffering and found that his conception of the condition that underlies the sort of suffering occasioned by the experience of pointless evil is different from that underlying the Perspective-Element and its idea of an 'independence of the world'.

For the Perspective-Element the context in which pointless evil is faced is that of the isolated individual who has to align his will to the will of the world, to achieve independence of the world. The perspective *sub specie aeternitatis* determines the self's reactions so that they are not subject to the way events go in the world. The perspective ultimately ensures that the self maintains *control* over its own responses so that the world's intrusion is kept at bay. And this provides a form of consolation. For no matter what happens in the world, the perspective on the whole of life, life seen *sub species aeternitatis*, holds the prospects that the self will not be ultimately violated, no matter how severe are the afflictions it suffers.

Bernanos sees the context in which pointless suffering occurs differently. He has no place for the aspirations to *control*, whether that control be directed outwardly at the world or inwardly at the self's own reactions in order to avoid the onslaught of the world. The orientation towards *control*, an orientation of particular fascination to the modern world whose degree of detachment from the spirit of childhood is acute, merely intensifies the condition of distress and misery and binds the self ever deeper in turmoil. Recalling a quotation from the beginning of this section, the aspiration to control is part of the very condition in which the modern world (and proponents of the Perspective-Element) see suffering.

> *All suffering is vain and intolerable for someone who lays the blame on the obstacle, like a child who beats with his fist an object on which he has hurt himself...* (Quoted in BH 57).

Suffering is vain also if it is seen in terms of the Perspective-Element, which is underlain by a fascination with control of the self to avoid external intrusions – that is what is epitomised in the urge for an *independence from the world*. For Bernanos an aspiration for such independence from the world is incompatible with the spirit of childhood and its capacity for pity and discernment of solidarity with others.

It is worth emphasising just how the Priest of Ambricourt's faith is irreducible to independence of the world. He has no aspiration to control his responses. When he confronts the prospect of his death, having had the stomach pains which haunted him since coming to the parish diagnosed as terminal cancer, the Priest of Ambricourt writes:

> *(J)ust when I need most strength, the knowledge of my weakness so entirely overcomes me, that I loose the thread of my paltry courage...* (J 236).

He confronts his death with terror, forgetting the reality of God and viewing death as an erasure of the self's existence (J233). Later, when he gains at least some composure he still feels that his death will be as small and awkward as his life. There is here no grand disengagement from the world but a desperate attachment to it. Moreover, he feels no confidence at all that he will be able to be independent, in the sense of being aloof from, his impending fate.

> *I shall be no better at dying than I am at controlling my life. I shall be just as clumsy and awkward...My courage shall be to know that I have none...* (J 237).

The spirit of childhood is exemplified in the stance of a victim, a victim at the mercy of God's grace. When the self aspires for an independence of the world it no longer possesses the character of a victim. Vulnerability and weakness are the conditions of the manifestation of Christ. For Bernanos, the weakness of fickle Man is such that only weakness can 'reach' him. The Priest of Ambricourt is not self-assured, he is vulnerable; he threatens no one. He does not appear as one self struggling for dominance over another. No one puts up their guard against him, he has no social prestige nor fluent manner to upstage anyone. His vulnerability enables him to achieve the capacity for genuine pity. Because he has no stake in the world of human hierarchies, hierarchies of individual popularity among men as much as of formal power, he is a barometer of what those in *angoisse*, in anguish and distress, have to confront about their own lives. His presence makes them confront themselves in their ultimate weakness, confront the reality of the illusions they cherish and hide behind.

In conclusion, the view of suffering presupposed by the Perspective-Element with its idea of an independence of the world is not intrinsic to self-renouncing belief. It is at best one possible conceptualisation of how suffering is understood in religious life. Moreover, it is one which Bernanos does not view as inherently self-renouncing.

Bernanos and the Anti-Consolation Element

The Anti-Consolation Element as distilled in Chapter 1 implies that self-renouncing belief in God is not held for any consolatory purpose(s). In this section I shall argue for the following conclusions. Firstly, the Anti-Consolation Element fails to give due regard to some contexts in religious life where some forms of consolation have a place. Secondly, the concept of consolation which the Anti-Consolation Element presupposes is over simplistic. Consolation is assumed to necessarily be some sort of gluttonous desire for what is distinctly and unambiguously self-directed. This involves the uncritical view that all forms of consolation involve a non-problematic split between what is an end-in-itself and what is an end-for-the-self. For Bernanos, some forms of consolation, principally related to the expression of some varieties of emotion, defy this sort of rigid division. The experience of them is actually an expression of the spirit of childhood.

Let us start with the first point, namely, the way the Anti-Consolation Element fails to give adequate regard to the context in religious *life* where consolation has a place.

Bernanos saw the role of the Catholic writer as the 'perpetuelle recherche de l'etre', which includes providing a portrayal of the place of evil in our lives (see J.C. Whitehouse, 'Teaching, Witness and Vision: Some Reflections on Bernanos' View of the Responsibilities of the Catholic Writer', *Romantic Review* 11, 1987, 99). In the *Journal* evil is found in the anguish and despair, the *angoisse* and the boredom that infects the parish, the desires of many characters (such as the Comtesse) for avoidance of meaningful contact with others, the lack of vocation in many of the clergy and their compromise with worldly standards of morality through sheer lack of interest in making any deep contact with others. The novel commences with reference to the Priest of Ambricourt's vocation to his parish. And in this there is a sense of foreboding. 'Mine is a parish like all the rest. They're all alike. Those of today, I mean' (J 5). In the very first paragraph we sense that something is wrong when the free-thinking comments of the Priest of Norenfontes arouse in other priests a sense of glee. In the Priest of Ambricourt they bring on such a deep sense of discouragement (J 5). He senses in the modern world a 'leprosy of boredom' (J 6). This condition has infected the notion of vocation in the priesthood itself.

> *Our superiors are no longer official optimists. Those who still profess the rule of hope, teach optimism only by force of habit, without believing in what they say* (J 6).

The state of boredom is reinforced in the novel by the atmosphere of the dank, rain infiltrated open country and the almost endless roads the priest has to travel on his bicycle to visit his far flung parishioners. A central theme of the *Journal d'un cure de campagne* is of the struggle of the Priest of Ambricourt against such a despairing condition.

What is the root of this condition of boredom? It is the directedness towards self and the intensely modern capacity for sustaining an individual life devoid of all sense of the reality of others. For Bernanos, that condition of the self where there is no communion with others is the most penetrating form of evil. But such evil, and the associated despair, is something Bernanos thinks most people are in one sense unconscious of. This is not to say they are unaware that something is wrong. But in their self-orientation they have no real feeling for where the problem lies.

But for one endowed with the spirit of childhood the condition of despair becomes more pronounced and more directly felt. It is the saint, endowed with the spirit of childhood, who most clearly discerns the barriers to solidarity that the world erects. The saint discerns the impurity of human affairs and the way individuals utilise one another, far more deeply than the average person. Consequently, the saint faces a far greater burden than the worldly man. Moreover, one endowed with the spirit of childhood lacks worldly self-assurance and does not proudly believe that his or her own self is strong enough to withstand the penetrating temptation to despair and boredom that threatens it and others. The Priest's inner strength seems to melt away in floods of tenderness and rising tears as he comes to confront his inability to give practical expression to his desire to foster solidarity among those he meets (J 10).

The novel includes many accounts of the Priest of Ambricourt's despair and his overwhelming sense of destruction (J 33). Elsewhere he writes in his diary of that '…indefinable exhaustion, as though my very soul were bleeding to death' (J 76).

On another night he makes a furious effort to pray,

> *Nothing…the night is entering into me by some inconceivable gap in my soul. I, myself, am the night* (J 91–92).

The first half of the *Journal* is filled with the Priest of Ambricourt's almost constant struggle against despair.

It is at this point that we discern one context in which some forms of consolation have a legitimate place in religious life. The spirit of childhood discerns the dark hopelessness which lack of solidarity with others and a self absorbed in its own interests to the exclusion of all else, ultimately brings about. As well as discernment, the spirit of childhood awaits in humility for, and aspires to, the consolations which God alone can provide through such intensification of solidarity with others; it knows that the self is unable to furnish them. Believing in God on the basis of hoped for consolation is, in this context, not only compatible with but is also *indicative* of the spirit of childhood itself.

To summarise this first conclusion, the Anti-Consolation Element fails to give adequate regard to some contexts in religious life where consolation has a place.

It is important to emphasise one point. That Bernanos' conception of self-renouncing faith has a place for belief in God on the basis of consolations does not mean that *all* forms of consolation are religiously legitimate. Bernanos is not at all ignorant of the way *some* forms of seeking for consolation derives from spiritual gluttony.

The novel is replete with examples on this front. The Priest of Ambricourt felt himself, at one stage, caught up in this:

> *Once... (my praying)...had an obstinate, imploring quality...then insistent, imperious even – yes, I would have liked to snatch His graces from Him...* (J 197).

There are also consolations derived from believing in God in order to secure a sense of one's own strength. The Priest also discerns he has fallen, at times, into this trap and had to be purged of it: 'God, I presumed upon my strength. You cast me off into despair as we fling a scarce-born animal into the water, tiny and blind' (J 125). There are also consolations which produce spiritual complacency – belief in God is held as part of the secure and familiar life to which the believer feels he or she can return to at any time (J 90).

Another form of getting consolation from belief in God that is incompatible with the spirit of childhood is that in which the self gains strength from actively portraying itself as a victim of injustice. The Priest talks of those victims of iniquity who are able to find in that knowledge that they have been hard done by, some basis of strength (J 247). But the Priest then writes humbly that *he* would hate to be a cause of another's sin (J 247). For Bernanos, Christ did not hold Himself as a victim of injustice because that would have been an indirect way of attributing sin to others. Bernanos also extends this idea to cases of seeing the world as an intrusion into the self's individual autonomy when suffering is experienced. Thriving on this picture of the *individual* self as the victim is a way of dismantling the solidarity with others that the spirit of childhood aspires to, as we have seen in the section on Bernanos and the Perspective-Element.

Turning now to the second line of argument noted at the beginning of this section. Consolation is not necessarily, as implied by the Anti-Consolation Element, something gluttonously and straightforwardly *self*-directed. Indeed, we have already noted how the *hope* for consolation can be something other than a matter of self-orientation. We shall now explore how the *experience of some forms of consolation is actually an expression of the spirit of childhood* and, further, how the division 'for self/not for self' breaks down in such contexts.

In this connection the most important example is Bernanos' conception of *joie* (joy). The Priest of Ambricourt writes in his journal: 'The mission of the Church is to discover lost joy' (J 230). Bernanosian joy is not what accompanies the worldly person's consumption of the things to be had from this world. Rather, it exemplifies the spirit of childhood in its delightful wonder at the realisation of the solidarity of the self with others. Joy is illustrated in some detail in the figure of Chantal in *La Joie*. Her spirit of tenderness and trust disturbs the other characters, provokes them to confide in her and exposes the futility and evasions in their lives. But Chantal's joy seems artificially infused into her and it is in the *Journal* that we find Bernanos' more mature concept of joy, a joy hard earned and reaped from much misery and pain.

After long periods of despair the Priest of Ambricourt comes to know that joy. Not long before his death he experiences friendship with the Comte's nephew, Olivier, before the latter's return to his regiment. Previously he had '...never dared to be young' (J 200), being unable to open his heart to others throughout his period at the seminary when he had to carry the shame of his background and his poverty. Bernanos describes one motorcycle trip which Olivier and the Priest make together. Now he realizes

> *...that youth is blessed...a risk that is also blessed...I knew that God did not wish me to die without knowing something of that risk...* (J 200).

The Comtesse, after her conversion by the Priest of Ambricourt, also comes to know joy. In the note she sends him before her death she writes:

> *...I wonder what you have done to me...Or rather, I no longer ask myself. All's well. I didn't think one could ever possibly be resigned...I'm not resigned, I'm* happy... (J 150).

This is an indication of the joy to be had in faith as perceived by Bernanos. Her joy is passed, for a time, to him. 'I know that never before have I experienced, or shall again, hours of such fullness' (J 152). He recalls his earlier words to her.

> *'Be at peace', I told her. And she knelt to receive that peace. May she keep it for ever. It will be I that gave it to her* (J 154).

This is not pride in his own accomplishment, but a delight that he has been the vehicle for her confession and can share with her deeply in that, through the joy he feels.

Such joy is *creative* at the level of the Church as a community. Through it the members of the Church sustain each other. Bernanos sees such joy as akin to the joy of parenthood. This is particularly clear in the scene when the Priest sees the dead body of the Comtesse laid out in the coffin. He realises that self-renunciation is essentially a reaching out to others:

> *I lifted the muslin from her face, and stroked her high, pure forehead, full of silence. And poor as I am, an insignificant little priest, looking upon this woman only yesterday so far my superior in age, birth, fortune, intellect, I still knew – yes, knew – what fatherhood means* (J 154–5).

For Bernanos the joy of fatherhood is, in its pure form, not something in which there is any tension of the form 'for self/not for self'. The Priest's joy is of this sort. Through it we discern his sense of communion with the Comtesse, despite their differences in rank and accomplishments, and his sense of wonder at their both being channels of the fearful care and terrible love of God.

For the Priest of Ambricourt such joy finally becomes the means to a form of compassion for himself. The wonder and awe at being himself the bearer of a love that attracts others away from their absorption with self is something he learns about the utility of his lack of self-assurance and 'awkwardness' (*maladresse*). He can say: 'I have always known that I possessed the spirit of poverty' (J 237).

The Prioress in *Dialogues de Carmelites* says:

> *(N)ever despise yourself. It is very difficult to despise ourselves without offending God. Self-contempt leads straight to despair* (DC 61).

Self-contempt leads to an *angoisse* which makes compassion for others impossible. At the end of the *Journal* the Priest of Ambricourt court can write:

> *The strange mistrust I had of myself, of my own being has flown... (I)f pride could die in us, the supreme grace would be to love oneself in all simplicity* (J 251).

Bernanosian joy makes possible a form of love where the 'self/not-self' division implied by the Anti-Consolation Element is inapplicable. The spirit of childhood, when it attains joy, is able to love the self and to treasure it, just as it can love and reach out to others.

A further example of an emotion which defies the split between the simple 'for self/not for self' division is *pitie* (pity or compassion) which has already been discussed in the section on Bernanos and the Perspective-Element.

The conclusions of this section are as follows. Firstly, the Anti-Consolation Element fails to give adequate regard to some *contexts in religious life* where some forms of consolation have a place. When this context is more fully appreciated we recognise how some consolation is readily compatible with self-renunciation of the form Bernanos calls 'the spirit of childhood'. Secondly, the concept of consolation which the Anti-Consolation Element presupposes is over simplistic. Some forms of consolation are associated with emotions found in the religious life (such as *joie* and *pitie*), emotions which are integrally linked to the concept of the spirit of childhood, and do not entail a non-problematic split between what is an end-in-itself and what is an end-for-the-self.

Bernanos and the Absoluteness-Element

According to the Absoluteness-Element discussed in Chapter 1, self-renouncing adherence to God is absolute; it is discontinuous from all relative ends, ends which ultimately pertain to the self. In this section I argue that the Absoluteness-Element is not intrinsic to Bernanos' model of self-renouncing faith. Indeed, on his view it is not inherently self-renouncing because the aspiration to hold belief in God as an end in itself can negate the spirit of childhood which forms the epitome of the Bernanosian form of self-renunciation.

Let us start with the words of the Abbe Menou-Segrais in Bernanos' *Sous le soleil de Satan*:

> *(E)ach one of us is alternatively borne towards the good, not by a thoughtful calculation of its advantages, but simply by a drive of the whole being, a pouring out of love, which turns suffering and renunciation into an object of desire... (or)...at other times tormented by a mysterious desire for self-degradation...an unfathomable hankering after evil...Evil, like good is loved for its own sake, and served too* (SS 232–3).

The important point to note is that Bernanos is here making a point about two forms of orientation in which things are held as ends in themselves but which are not authentically religious. Loving evil as an end in itself is one example. Turning suffering and renunciation into objects of desire is another. Significantly, both these orientations are part of the same continuum of oscillation that characterises human nature untouched by grace.

It might be insisted that having renunciation as an object of desire – an end in itself – is surely authentically religious when this is done out of belief in God. But Bernanos came to oppose such a view. His opposition is expressed in the dissatisfaction he felt with the portrait of sanctity associated with the character Donissan of *Sous le soleil de Satan* and the way he felt the Priest of Ambricourt comprised a truer picture of self-renouncing faith.

In the earlier novel, Donissan, when told by his spiritual superior that 'When God awaits you, you must ascend, either that or lose yourself' (SS

111), sets about making himself a saint by will-power. In undertaking additional acts of devotion Donissan finds a sense of joy. This he treats with suspicion and seeks to root out with agonizing self-flagellation.

> *He was at that stage of paroxysm where disappointed love discovers new strength to destroy...he was punishing 'this body of death' from which the apostle prayed to be released, but then the temptation entered more deeply into his soul and he hated himself, completely. Like a man who cannot outlive his dream, he hated himself* (SS 132).

That degree of disinterestedness that is the ultimate pinnacle of holding belief in God as an end in itself is unattainable. But in Donissan's assertive striving for such a state, the recognition of its unattainability brings self-hatred that threatens to seal him off from others.

An example of such extreme isolation from others is the case of the Comtesse, whom the Priest of Ambricourt confronts and converts. But in her initial move to accept such conversion she tries to substitute belief in God as an end in itself for what was formerly an end in her life, namely the memory of her dead son. That that was formerly such an absolute in her life is suggested by her self-sacrificing dedication to it. She says: 'Nothing, either in this world or the next, can separate us from what we've loved more than ourselves...more than getting into heaven' (J 141). The Comtesse affirms that, if there existed a place beyond the reach of God, a place to which, having recovered her dead son, she could take him, then she would do so (J 146).

In the novel we find that this absolute attachment has precluded her from any interest in any other human relationship. She has lived for years in a state of indifference towards her daughter and her husband. Her manner is cold, dismissive and curt. For Bernanos, her isolation from others is a manifestation of evil. Evil, for him, is a vast yearning for the void, for emptiness (J 125).

After a long drawn out confrontation the Priest of Ambricourt brings her to a realisation of the insufficiency of her condition. She is finally brought to acknowledge God. But she is initially inspired to try to make her faith conform to the pattern of her apostasy, to make it involve a total rejection of all relative ends that pertain to the self and an attachment to faith in God as an end in itself. In a dramatic gesture that expresses this stance she throws the medallion containing some hair from her dead child into the fire (J 148). The Priest of Ambricourt condemns her daring and tells her that God wants us to be merciful with ourselves. The aspiration to disengage belief in God from all relative ends, such as those which pertain to the self's need for consolation in sorrow and through fidelity to attachments to other human beings, is neither required nor is it authentically religious.

The aspiration towards the capacity to hold belief in God as an end in itself, devoid of all relative ends pertaining to the self, can negate the spirit of childhood which forms the epitome of the Bernanosian form of self-renunciation. In the *Journal d'un cure de campagne* Bernanos is deliberately reacting against a portrait of virtue which emphasises the self's own capacity to attain a belief in God detached from all other, relative, ends. The novel contains an explicit resistance to the heroic view of virtue found in Paul Claudel.

The Priest of Ambricourt is presented as one who never presumes to think he has in himself the capacity to will belief in God as an end in itself. His sense of 'the powerlessness of us all' is what makes him so able to feel compassion for others. For example, in the face of death, he cannot resist the temptation to weep and to wish in desperation that his life were not to pass away.

> *...I shall be no better at dying than I am at controlling my life...My courage shall be to know that I have none* (J 237).

This is not the powerful saint who can throw aside concern for any relative ends and give himself absolutely to God. Bernanos' view of innocence derived from an awareness that an overemphasis on the ability to resist the need for any relative ends can lead to a reliance on the self's own strength which for him is incompatible with the innocence associated with the spirit of childhood.

The spirit of childhood is a mode of being that is characterised by the lack of worldly self-assurance. Bernanos wanted to show how aspiring for one's belief in God to be something devoid of all concern for relative ends is indicative of a concern with the self's own accomplishments. As soon as renunciation is presented as a project which the self seeks to master and control then there arises an obstruction to a genuine solidarity with others. Such a self can so often, through involvement with its own projects and concerns, fail to detect the fragility and the needs of others. It can also fail to inspire in them the capacity to confess and identify to themselves their fragilities and weaknesses.

Self-renouncing faith, far from being an absolute in the way presupposed by the Absoluteness-Element was seen by Bernanos as having an end. That end is a form of solidarity with others and with God. Bernanos sometimes talks of this solidarity as a form of *joy*. 'The mission of the Church', says the Priest of Ambricourt, 'is to rediscover the source of lost joy' (J 230). This joy is seen as both a *communal wholeness* and an *individual self-enrichment* that cannot be detached from one another. That self-enrichment and communal wholeness is something which can be seen, at times, to be an end separable from belief in God.

We can summarise the findings of this section as follows. The Absoluteness-Element is neither inherently self-renouncing nor intrinsic to self-renouncing faith. For the aspiration that one's faith be characterised by such total disinterestedness is, for Bernanos, incompatible with what is for him the paradigm of self-renouncing faith, namely, the spirit of childhood.

Bernanos and the Anti-Metaphysical Element

It will be recalled from the discussion in Chapter 1 that the Anti-Metaphysical Element implies that belief in God is self-orientated when it is involves a contrast between the world that we are part of and another realm beneath, behind, beyond or to come later. Self-renouncing faith, supposedly, is a faith able to accept the vicissitudes of this world without requiring resort to the security of positing a metaphysical realm or Being. It can accept what D.Z. Phillips calls the 'inherent pointlessness in things'. And a faith which is *unable* to accept things without bringing in the metaphysical is a faith that is self-orientated. Is such a conception implied in Bernanos' view of self-renouncing faith?

As we have seen in previous sections of this chapter, there are several concepts which have traditionally been assumed to involve a metaphysical dimension which in Bernanos are not straightforwardly linked to metaphysics. Thus the concept 'eternal life' was found not to be essentially about life of unlimited duration but about a life in solidarity with others. A further example centres on my rejection of the idea of the application to Bernanos of the idea of vicarious substitution as a matter of the metaphysical transfer of emotional states.

Bernanos' interpretation of the doctrine of hell seems also to depart from the traditional metaphysical account of it as some sort of 'place'. Hell, the Priest of Ambricourt tells the Comtesse, is not to be understood as '...penal servitude for eternity' (J 139). Such a view is the product of a society and an attitude which sees things in terms of what is most to its convenience. '(M)oralists, even philosophers refuse to see anything but the criminal... they form no idea of essential evil, that vast yearning for the void' (J 125). Thus they fail to see, in a way that the spirit of childhood can see, that the criminal is often only a victim crushed by fate.

> *To judge us by what we call our actions is probably as futile as to judge us by our dreams* (J 76).

But to see in this way is inconvenient for the those who cling to worldly standards. They find it easier to label such individuals 'enemies of society' and spare themselves the trouble of understanding them (J 140). But for the spirit of childhood, hell is not some convenient realm which serves to comfort those in authority in this world, to make them feel at ease by giving Divine approval to their convenient labelling of unfortunates they cannot be bothered with. Hell, says the Priest of Ambricourt, is '...not to love any more' (J 140), the state in which there is an absolute incapacity to respond to others.

Despite the above example I now wish to argue that Bernanos does not have any reason for implying that there is a necessary link between bringing in metaphysical dimensions and the urge for self-security and self-orientation.

The reason for this is conceptually connected to the concept of the 'spirit of childhood' as it occurs in his thought. For the notion of the 'spirit of childhood' is in Bernanos something quite compatible with many forms of consolation for the self. As we have seen in the section on the Anti-Consolation Element, despair that affects individuals, and especially the saint, is often so intense that the hope provided by consolation is what the self directs itself to in order to maintain an orientation towards God and the spiritual life.

Indeed it is indicative of the spirit of childhood that it regards itself at the mercy of God's grace to provide it with the consolation it needs, but cannot of itself provide or muster, to remain faithful to the religious life. For those who are poor and without hope then the spirit of childhood does not begrudge them the consolation of a 'dream of splendour' to make their suffering a little less intolerable. Bernanos would extend this to the idea of a metaphysical realm, even where that is brought in to provide consolation.

The above point needs qualification. Bernanos would not allow that all consolation is compatible with self-renouncing faith. He believed that there is a tendency among the privileged sections of society to seek a theodicy as soon as there is any sign that their privileged way of life is felt to be under threat. The Comtesse, for example, is an individual who believes in a metaphysical God, and carries out the religious devotions of the Church, in order to spite such a God. Hers is a form of self-orientation that *requires* a metaphysical God. That is to say, her self-affirming and spiteful rejection of God is only intelligible in terms of the conception of a God that belongs to another realm, looking into, but unaffected by, what goes on in this one. Here, recourse to the metaphysical is an expression of self-orientation.

Is the idea of God as metaphysical *necessarily* something that involves a sense of a bond to something which promotes self-security? For Bernanos the idea of God as metaphysical is not primarily that of a *mechanistic* controller from on high. Sister Constance in *Dialogue des Carmelites* says:

> *(I)t does seem to me to be less sad not to believe in Him at all, than to believe that He is some sort of mechanic, geometer or doctor...God is not a mechanic, a schoolmaster with his cane, nor a judge with his scales. If He were, then we*

> *should have to believe that He will take counsel of what are called solemn, level-headed sensible people, and that is a silly idea... (P)eople of that kind have always thought the saints were mad...* (DC 64).

As we have seen, Bernanos stresses that the essential act of God in intervening in the world is not with thunderbolts and displays of worldly power. It is in suffering. The God that Christians worship is a God that has assumed human form and has suffered. The power of God affects us not by physical coercion but by penetrating our psychological condition and calling forth our sympathies, addressing our natural condition of seeking out bonds of solidarity with others. Bernanos presents, in effect, a challenge to the simplistic conception of a metaphysical controller presupposed by the Anti-Metaphysical Element. For Bernanos, God is not metaphysical in the sense of being in charge of a vast array of levers nor because He is an excellent computer programmer. The influence of God is felt in the world and changes its constitution all the same. It operates in the way that the spirit of childhood effects change in the world.

The Anti-Metaphysical Element holds that there is something inherently self-renouncing about a faith which has no sense of a bond to anything metaphysical, a faith which is able to accept the 'inherent pointlessness in things'. How does Bernanos stand in relation to this facet of the Anti-Metaphysical Element? Bernanos offers us a model of faith in which this is questioned. The spirit of childhood is characterised by its capacity to be 'bonded' to others, including a metaphysical God. The spirit of childhood does not have a stance towards the world of *accepting* pointlessness. Its nature is orientated to finding and displaying, or calling forth, in others a certain sense of there being a point in things. It provides testimony to the strength and the reality of solidarity with others and the need to abandon the view of the self as discrete, unattached and struggling in isolation.

In Chapter 1 it was noted that the Anti-Metaphysical Element implied that self-renouncing faith involves accepting the cessation of the self at death. Does this hold true of Bernanos' conception of self-renunciation?

In Bernanos we find that not ruling out a continuation of the self after death is an expression of the nature of the attachment the spirit of childhood has to the claims and relationships of this world. For the spirit of childhood does not dismiss all the claims of this world outright, as has been previously argued. The nature of its attachments to this world is sometimes expressed in a form of valuing them that involves not consigning them to a final temporal cessation. Bernanos does not see accepting finality as self-renouncing. He sees it as seeking the void of emptiness, something human beings in despair can yearn for as offering a vision (in fact an illusory one) of consolation. But it is consolation only to the aloof self wishing to reject the claims of solidarity with others.

Any emphasis on accepting the finality of death is incompatible with Bernanos' conception of the spirit of childhood. Though the Priest of Ambricourt shows no hankering to survive death and forms no clear view of what death entails, this is far from an acceptance of the finality of death. After his conversion of the Comtesse, he looks back at the incident after learning of her subsequent death. He recalls her struggle to overcome her resentment of God:

> *The memory of her struggle before my eyes, that fight for eternal life from which she emerged exhausted victorious, became painfully vivid... 'Be at peace', I told her. And she knelt to receive this peace. May she keep it forever. It will be I that gave it to her* (J 154).

Here we learn that the Comtesse's struggle for eternal life was victorious. The priest had witnessed her victory before his eyes. She has received eternal life before the Priest of Ambricourt or anyone else knew she was going to die. Therefore it seems clear that 'eternal life' here is not a matter of unending life. What eternal life means is the life of solidarity with others that the Church aspires to mediate, as elucidated in previous sections.

Does not this mean that accepting the finality of death is intrinsic to self-renouncing faith as conceived by Bernanos? Far from it. The fact that eternal life is to be found in this present life does not mean it is confined to this life.

The Priest of Ambricourt, endowed with the spirit of childhood, looks back to the scene in which he and the Comtesse had finally been reconciled. Death, the 'Gulf of All Light' as he calls it, has now separated them. But in his genuine love and *pitie* for her he wants that peace which she finally achieved, that peace mediated through him, a sickly, insignificant priest, to be with her for ever. He wants it to survive the gulf, the frightening disconnection from all earthly attachments that death is perceived to be. The point here is that the character of his stance towards the world does not think in terms of finality.

It could be objected that his wish that the peace he gave her will be with her 'for ever' is a mere form of poetic expression. In reply, we can see it is more than this when we consider the Priest of Ambricourt's attitude to his own death when he realises that he is to die. His response is totally different from one of acceptance of finality.

The spirit of childhood does not have in itself the confidence in its own strength to be able to aspire to detach itself from all the bonds of solidarity with others that holding out death as a finality entails. Considering his imminent death the Priest writes:

> *It has just occurred to me how my agony was that of a cruel, sudden disappointment...I shall be no better at dying than I am at controlling my life* (J 237).

The spirit of childhood cannot aspire to stoic indifference (J 249). 'I can understand how a man, sure of himself and his courage, might wish to make of his death a perfect end... (but)...my death will be what it can be, and nothing more' (J 249). He realises his part is not to be daring or 'defiant' (J 237). The Priest of Ambricourt is not a man sure of himself. He is not sure enough of himself and his own powers to aspire to accepting death as a 'perfect end', as a complete termination.

Further, it is integral to the spirit of childhood and the character of its bonds in solidarity to others that he cannot aspire to coldlessly embrace finality. For he writes:

> *...to a true lover, the halting confession of his beloved is more dear than the most beautiful poem. And when you come to think of it, such a comparison should offend no one, for human agony is beyond all an act of love* (J 249).

The point here is that the confession of his attachments to others in this life (finally achieved after the struggle and despair which characterises most of the novel except the final sections) means he cannot aspire to put them aside and seek out the aspiration to accept the finality of death. Because of those attachments, and the fact that death will as a result be ever more one of 'agony', it is integral to his love that he endure that. For human agony is beyond all an act of love. What he finally realises is that he does love the world: 'I was crossing the world without seeing...without even admitting it to (myself)' (J 234).

Experiencing the agony of being unable to accept finality while yet having to face death – death as the disconnection from all that he finally realises he loves – is integral to the nature of his love. The Priest dies having finally come to love himself and the world – he does not die having achieved an independence of the world (as posited by the Perspective-Element).

A further reason that Bernanos' view of self-renouncing faith does not incorporate an acceptance of the finality of death is that he is suspicious of its spiritual value. His characterisation of the spirit of childhood involves a deep sense of solidarity with others. Evil is the opposite of that solidarity. Evil is, in the words of the Priest of Ambricourt, the vast yearning for the void (J 125). The Comtesse is portrayed as one who, as a result of suffering, has been reduced to a terrible concentration of self. She has become indifferent to her family, to other people and to God, and clings desperately to the memory of her dead son in a way that, as the Priest shows her, is actually an infidelity to that memory. Her love of what is in essence her self-created image of her dead child is but a form of obsessive self-orientation that is a cancerous growth preventing any enduring solidarity for both the self and the Church. And the underlying decay of her marriage is as much due to her as to the sexual infidelities of her husband that her indifference has chosen to overlook. In her indifference she desires a place where she can be free of all her family, including God, and left alone with the image of her departed son. The Priest of Ambricourt sees this as a yearning for the finality of all attachments to this world. Such a stance towards the world is self-orientated, not self-renouncing.

It might be assumed from the above discussion that the Bernanosian idea of the spirit of childhood expects to survive death. This would be too imprecise a way of putting it. The spirit of childhood does not positively aspire to such survival. The point is that it does not accept finality – its stance towards the world is distorted if we think of it as something endowed with such a confident, defiant grasp of things in the face of death.

Another issue discussed in Chapter 1 was the Wittgensteinian position's assumption that conceiving of God as metaphysical is to reduce His reality to that of a worldly power. God's power is not to be understood as akin to physical power, able to vie with other physical powers, to overcome them and so effect a change in the order of things in the world.

This latter view is not replicated in Bernanos' conception of self-renouncing belief in God. For Bernanos, God is a metaphysical Being, His power is one which vies with other worldly forms of power but without itself being worldly. How are we to understand this?

The best way to understand Bernanos' conception of the power of God in the *Journal* is to understand the nature of the strength of the Priest of Ambricourt, a strength that is present through his powerlessness. This is not to say that God's power and the Priest's strength are in every way comparable. The Priest, for example, can talk of his own impracticality and also his inability as a leader (J 122). Now it makes little sense to talk of God as impractical or as a poor leader.

Nevertheless, there is a sense connected with the character of the spirit of childhood in which the Priest's strength-in-weakness is illustrative of God's power. Bernanos intended the Priest of Ambricourt's spirit of poverty to be linked to God's revelation in Christ. At one point in the novel the Priest realises that he is '…never to be torn from that eternal place – that I remain the prisoner of His Agony in the Garden' (J 174). Bernanos intended the Priest's strength-in-weakness to be a portrayal of Christ's power.

In the *Journal* the revelation of Christ is contrasted with worldly expectation. Torcy says that the poor were expecting something different. They were hoping for the relief from distress (J 50). How are we to understand the strength-in-weakness of the Priest of Ambricourt? This has already been touched upon in the discussion on the power of *pitie* (see the section on Bernanos and the Perspective-Element) but some restatement and elaboration is appropriate here. The Priest of Ambricourt, not long before his death, comes to realise that his strength is '...the strength of weaklings and of children'. One place in the novel where the meaning of this is made clearer is in a speech by Torcy to the Priest of Ambricourt. Torcy attacks those social reformers who think the Christian goal is the eradication of poverty. For among the poor, just as among the rich, the fascination for worldly power has as great a hold: the drunkard in the gutter is drunk with the same dreams as the most ambitious politician (J 56). To such reformers his message is:

> *...you'd do better to look at yourselves in the mirror of want, for poverty is the image of your own fundamental illusion. Poverty is the emptiness in your hearts and in your hands...* (J 56).

Torcy is not saying that social deprivation is not to be countered. What he is attacking is the equating of poverty with social deprivation. Poverty is the 'emptiness in your hearts and in your hands'. Poverty is the emptiness which cannot be ultimately filled by those whose law of life is counting debts (J 56). Bringing to bear the powerful resources of this world will not cure poverty as conceived of by Bernanos.

How then is this poverty to be cured? Torcy elsewhere talks of the power of *pitie*. In worldly terms *pitie* is often thought of as something snivelling and silly, whereas in reality it is something powerful and devouring. In the Priest of Ambricourt's confrontation with the Comtesse we see something of this *pitie* when he identifies with the suffering that has led to her sin of hating everyone including God:

> *As I listened sadness overwhelmed me, indefinable sadness against which I felt quite powerless...Suddenly I could feel a tear on my cheek...She watched this tear fall* (J 141).

It is to this awkward, impractical, sickly and simple priest that the Comtesse, who had lived for many years in that horrible and least human form of despair, makes her confession (J 144).

The very weakness of the Priest of Ambricourt becomes the source of his strength in converting the Comtesse. His *pitie* (pity/compassion) is non-judgemental and humble. She is from a prosperous background, used to social encounters in which one faction is seeking social dominance over another. But the Priest of Ambricourt is one from outside the entire tradition of aspiring to enhanced social standing or fluency in command of human beings. He is seeking no competitive advantage. And this makes it easier for her to confess. A point made in Rousseau's *Confessions* makes this clearer:

> *As soon as we get into the habit of measuring ourselves against each other, and moving outside of ourselves in order to be sure of getting the first and best place, then it is impossible not to dislike everything that surpasses us, everything that makes us smaller, everything that hems us in...* (quoted in Paul Zweig, *The Heresy of Self-Love. A Study of Subversive Individualism*, Princeton: Princeton University Press, 1980, 153).

The point here is that there is something about the psychological make-up of the self-orientated subject that makes it resist that which it perceives as a competitor.

This helps us see the relevance of the Priest of Ambricourt's non-judgmental approach and his refusal to blame the Comtesse. In fact he excuses her:

> *Oh, madame, nobody can see in advance what one bad thought may have as its consequence...The seeds of good and evil are everywhere... (H)ow closely we are bound to one another in good and evil...* (J 142).

This passage reinforces the picture of humanity as bound to one another in solidarity. But it is the fact of the Priest of Ambricourt's poverty, in the sense outlined above by Torcy, that effects the first step for the Comtesse's confession.

God conceived of as a supernatural power, stronger than all earthly powers, would be irrelevant to a conversion, to a change in the heart of Man, such as is effected in the Comtesse. The challenge she throws to such a God is that He has broken her already and that there is nothing more He can do to her (J 139). For Bernanos, only the 'power' of the spirit of poverty (which is equated to the spirit of childhood (J 237)) can overpower the most crippling forms of despair.

It is worth pointing out that Bernanos does not think the exercise of such 'power' as exemplified in the spirit of poverty is at all a rare phenomena. He believed that its manifestation percolates and supports a great deal of human life. This is well illustrated in his portrayal of the working class concubine of Dufrety, the former seminary acquaintance of the Priest of Ambricourt, who has abandoned the priesthood in order to 'find himself'. He masquerades as a writer but really lives off the money of his hard working lover who nurses him in his illness. The Priest of Ambricourt discerns in her, despite her lack of formal contact with the Church, the spirit of poverty. In her he detects the kindness which soothes drunkards, lulls babes, argues with relentless tradesmen, beseeches bailiffs and comforts the dying. Her voice is the 'voice of the working woman', the voice which holds out against the miseries of the world (J 244).

Contrary to the Anti-Metaphysical Element, Bernanos thinks that the power exemplified in the spirit of poverty, which is illustrative of God's power, *does* vie with worldly power.

Bernanos seems to imply a number of ways in which it does this. Firstly, as in the case of the 'voice of the working woman' mentioned in the last quotation, it does so by making possible, certain settled patterns of human life: by controlling drunkards, soothing children or the dying, and curbing the zeal of bailiffs anxious for the repossession of property. These kinds of interventions in the world are capable of redirecting or deflecting the activities of worldly forms of power which are centred, in Torcy's words, on 'debits and credits'.

Secondly, the spirit of poverty shows to the confident men of the world the 'mirror of want', their own 'fundamental illusion' and the 'emptiness in your hearts' (J 56). For Bernanos, self-orientation ultimately can only lead to emptiness, to a void. It is the spirit of poverty that points to a sense of our solidarity with others and enables the grace of God to be channelled. And it is only to this form of power, this non-competitive power-in-weakness, that the human psyche is ultimately receptive.

It might be objected that the above account of the power of God hardly requires God to be a concept with any kind of metaphysical reality. This objection fails to accommodate the subtle way in which metaphysical conceptions are integrated with Bernanosian self-renouncing faith. It will be recalled that we earlier stressed the manner in which the Comtesse *required* a metaphysical God in order for her rebellion and rejection of religious claims to be of the most intense variety. The same is true of the character of her faith after her conversion by the Priest of Ambricourt. The change in the orientation of her allegiances is achieved by *both* clarity about the nature of her own immersion in hate and a vague referential directedness to a reality outside her own self.

This latter directedness, thought vague, is also in one crucial respect specific. For not just anything external to the self will do. The role of 'God' as in this context is to inspire something at and beyond the very extent of imaginability. This is essential to conveying a sense of the maximal contrast achievable. Such a contrast could not be psychologically 'grasped' if the reality of God were merely conceived to be something embedded in tradition or in the categories of a language-game. The concept of the metaphysical God is alone capable of imbuing the Comtesse with a sense of a reality which is truly *other* than herself.

None of this should be taken as a denial of the fact that the idea of a metaphysical God is something with a long tradition which has endowed it with its own 'technical' vocabulary which may in part colour the way it is used in religious life. However, that is not a basis for trying to put aside the way that the notion of a metaphysical God has a genuine role in inspiring and encapsulating a sense of something distinct from the self and outside the arena of conventional orders of power and dominance. God as metaphysical is a way of conveying a sense of something as dominant over the self. But the nature of that dominance is not at all comparable to worldly power over another person.

In summary, we can note the following points that have emerged in this section. Bernanos shows us how a metaphysical conception of God need not be seen as a crude, mechanistic account motivated by desire for self-security. The spirit of childhood as exemplified in the Priest of Ambricourt does not involve an acceptance of the finality of death for three reasons. Firstly, it does not aspire to have the capacity for such a defiant, confident acceptance. Secondly, the character of its love and attachments to this world needs to be properly appreciated. In the case of the Priest of Ambricourt we see that experiencing the agony of being unable to accept finality while yet having to face death – death as the disconnection from all that he finally realises he loves – is integral to the nature of his love. Thirdly, Bernanos' has no means of distinguishing the acceptance of the finality of death from his view of Evil as a yearning for the void.

A final point to note is that Bernanos can conceive of God as metaphysical and having a power which vies with worldly powers but without ending up with a crude worldly conception.

All the latter points of summary suggest that the Anti-Metaphysical Element outlined in Chapter 1 is inadequate if applied to some conceptions of self-renouncing faith.

Conclusions

The Bernanosian model of self-renouncing faith as characterised by the 'spirit of childhood' is a fundamentally different model from that of the Wittgensteinian position, as outlined in Chapter 1.

(1) There is no Absoluteness-Element in Bernanos. The Absoluteness-Element is not intrinsic to self-renouncing faith. Nor is it inherently self-renouncing. For the aspiration that one's faith be characterised by such total disinterestedness is, for Bernanos, incompatible with what is for him the paradigm of self-renouncing faith, namely, the spirit of childhood.

(2) The Anti-Consolation Element is not intrinsic to Bernanos' view of self-renouncing faith. Firstly, the Anti-Consolation Element fails to give adequate regard to some *contexts in religious life* where some forms of

consolation have a place. When this is more fully appreciated we recognise how some consolation is readily compatible with self-renunciation of the form Bernanos calls 'the spirit of childhood'. Secondly, the concept of consolation which the Anti-Consolation Element presupposes is over simplistic. Some forms of consolation are associated with emotions found in the religious life (such as *joie* and *pitie*), emotions which are integrally linked to the concept of the spirit of childhood, and do not entail a non-problematic split between what is an end-in-itself and what is an end-for-the-self.

(3) There is no Unreflectiveness-Element in Bernanos' view of self-renouncing belief in God. The spirit of childhood is something deemed to be quite compatible with reflection and even doubt.

(4) The Perspective-Element is neither intrinsic to self-renouncing faith nor inherently self-renouncing, in Bernanos. For him suffering is not something that should be seen as confronting the isolated individual, who then seeks to control its responses to achieve an independence of the world. The spirit of childhood is incompatible with that.

(5) The Anti-Metaphysical Element is not intrinsic to self-renouncing faith in Bernanos. He does not imply that metaphysical elements in faith automatically renders that a matter of self-orientation. Firstly, the spirit of childhood is quite compatible with believing on the basis of anticipated consolation in *some* contexts. Secondly, he has a view of God as metaphysical which is more subtle than that of a simplistic idea of a kind of mechanic or a puller of levers, even though it is a God somehow beyond the world. Thirdly, the idea of aspiring to a bond to a metaphysical God is quite in keeping with the spirit of childhood, which is essentially a conception in which the (non-individualistic, non-solipsistic) self outreaches to others, including a metaphysical God.

Note

1. Quotations from this work will be taken from the excellent translation by Pamela Morris, details of which are given in the bibliography.

3 Roots of the Wittgensteinian Position: The *Weltbild* of Self-Concern

Introduction

The last chapter demonstrated that there can be authentic conceptions of self-renouncing faith that are different from that presupposed by the Wittgensteinian position outlined in Chapter 1. In this chapter I shall explore the roots of the Wittgensteinian position. My objective is directed at drawing attention to the type of *Weltbild* that underlies the Wittgensteinian theses and gives them 'life'. My use of the term *Weltbild* refers to a fundamental attitude towards existence. The main task will be to show how this *Weltbild* presupposes a distinct understanding of the *predicament of the self*. In the next chapter we shall be able to draw on this analysis to explain why the Wittgensteinian position fails to encompass the model of self-renouncing faith found in the case-study in Chapter 2.

It is important to emphasise that this chapter will not take the form of a conventional history of ideas. Reference to historical precedents will not be as systematic and comprehensive as would be required if the object was merely to find a trail of ideas going back through different historical schools of philosophy. For the aim is to make clear the *character* of the *Weltbild* of self-concern. The full appreciation of this character requires us to see it in relation to its situatedness in cultural life, that is, in the context of some aspects of the spirit of the times in Continental Europe in the late nineteenth and early twentieth centuries. Tracing minute details relating to influences will not be important. Indeed, I shall bring in references to thinkers where no direct documentary link to Wittgenstein can be demonstrated but when this elucidates further the situation in life of the *Weltbild* of self-concern.

The next section will set out the main facets of the *Weltbild* of self-concern and its sense of the predicament confronting the self. Subsequent sections will examine how the *Weltbild* underlies each of the Wittgensteinian elements elucidated in Chapter 1.

Cultural Context of the *Weltbild* of Self-Concern

Although this chapter will range quite widely, the following individuals will receive direct attention as *exemplifying* the *Weltbild* of self-concern: Tolstoy, Georg Trakl, Rainer Maria Rilke, Hugo von Hofmannsthal and Ralph Waldo Emerson. Among the influences that *led* Wittgenstein to this *Weltbild* are Schopenhauer, Karl Kraus, Otto Weininger and Oswald Spengler.

The latter are specifically mentioned as influences by Wittgenstein (CV 19). But what of the names on the former list? What justification is there for choosing to look at them? The answer is that they yield a form of self-concern that shows us more clearly the character of the independence of the world that underlies Wittgenstein's orientation to life. This will become clearer as we explore how each of the Wittgensteinian elements can be understood as rooted in this sort of self-concern.

However, the names on the former list do have, to varying degrees, connections with the cultural milieu in which Wittgenstein lived or had contact with. Wittgenstein is known to have read Tolstoy and, for part of his life, to have practised a form of Tolstoyan Christianity as Monk has noted (LWDG 116). Wittgenstein never met Trakl, though was on his way to meet him before the latter committed suicide. Trakl and Rilke were among the writers Wittgenstein donated money to through Ludwig von Ficker. Wittgenstein described Trakl's poetry in a letter to von Ficker as having the '...tone of true genius' (ed. C.G. Luckhardt, *Wittgenstein: Sources and Perspectives*, Sussex: Harvester Press, 1979, 88). In view of what Monk says about Wittgenstein's reverence for genius, his obsessive honesty and propensity to avoid inflated descriptions of people, we should take this as an indication of Wittgenstein's sympathy for what he had seen of Trakl's work. Rilke was one of the few modern poets that Wittgenstein admired (see LWDG 108). Rilke, Trakl and Hofmannsthal were Austrian – the latter two closely associated with Vienna, as was Wittgenstein. Wittgenstein is known to have read Emerson, as both Monk (LWDG 121) and Brian McGuinness record (*Wittgenstein: A Life*, Berkeley: University of California Press, 1988, 224). (Monk denies all similarity between Wittgenstein and Emerson. Although I concede there are major differences, I hope to show they do share a comparable self-concern. Emerson was influenced by Kant and his stress on individual consciousness is largely due to German Idealism, rather than American individualism).

Some connections to Kierkegaard will also be brought in from time to time. In spirit, Kierkegaard was close to exemplars of what I am calling the *Weltbild* of self-concern. Rilke was very familiar with his work and actually went so far as to learn Danish in order to read him in the original. Writing to Drury Wittgenstein said that 'Kierkegaard was by far the most profound thinker of the last century. Kierkegaard was a saint' (Correspondence published in *Acta Philosophica Fennica* 28, 1976, 1–6).

i. The Perceived Predicament of the Self

The *Weltbild* of self-concern is one reaction to a perceived sense of the predicament of the self arising among a wide circle of intellectuals during the latter part of the last century.

The basic aspect of this predicament is the sense of the *isolated self facing a vast, impersonal universe*. It could be said that something of this sense of isolation goes back at least as far as Pascal, one of the first Christian thinkers operating under a new cosmology. But Romanticism had, as Charles Taylor suggests, helped to reintegrate something of an older, meaningful order by its view of a great current of life running through everything, including ourselves (SOS 416). By the end of the nineteenth century, though, the scientific mentality had affirmed more poignantly the sense of an external universe as impersonal, empty of apparent value, and unconnected to the interests of human society. Thus Tolstoy's Ivan Illych, reflecting the author's own personal crisis, finds himself facing death '...all alone on the brink of an abyss, with no one who understood or pitied him' (L. Tolstoy, *The Raid and Other Stories*, trans. L. & A. Maude, Oxford: Oxford University Press, 1982, 255). Here is a perception of a universe devoid of any transcendent reality. In his book *The Savage God: A Study in Suicide* Alvarez talks of the '...collapse of the whole framework of values by which experience was traditionally ordered and judged – religion, politics, national and cultural traditions' (SG 181). Vienna, previously an important seat of government, now experienced the weakening of an empire and a whole order of society. Tolstoy, a well travelled Russian aristocrat, experienced the same sense of imminent decay within the institutions of his own country, as did so many countrymen of his class.

One facet of this sense of isolation of self is found in Hofmannsthal. Hofmannsthal experienced a sense of the mutability of personal identity in the face of the relentless forward movement of history and external circumstance:

> *This is a thing that mocks the deepest mind*
> *And far too terrifying for lament:*
> *That all flows by us, leaving us behind.*
> *And that unhindered my own self could flow*
> *Out of a little child whom now I find*
> *Remote as a dumb dog, and scarcely know* (HHP 26).

Here the self 'flows out' and is carried, by a process over which it has no control. Moreover, such a process leads it to a state experienced as discontinuous and unrelatable to what it was previously. F.M. Sharp in *The Poet's Madness: A Reading of Georg Trakl* highlights how the sense of the self's reality, as traditionally conceived, became questionable and refers to the Viennese '...age in which a vital intellectual current professed a conception of inner and outer reality with extremely malleable boundaries' (PM 65). Ernst Mach's views that the self has no reality independent of sensations in the mind had contributed to this intellectual climate in Austria, as Freud was to do slightly later.

Related to this was the perception of the self as a victim of contingencies. F.W. van Heerikhuizen explores this theme in Rilke's work and concludes that 'the feeling of being at the mercy of things is...typical of Rilke' (RMRH 22). Trakl, whose whole life was one of drug taking in order to overcome the sense of a hostile world, is a further case of an individual who felt powerless: 'It is such an indescribable disaster when one's world breaks apart' (PM 31). This is from one of a stream of letters to his friend, the publisher von Ficker, expressing his sense of being inwardly broken.

A widespread view of history exacerbated the feeling of being at the mercy of contingencies. Tolstoy's *War and Peace* (1869) presented history as a vast and uncontrolled flow of events. It is something outside the command of individuals, including even the most powerful leaders such as Napoleon Buonaparte. Spengler, who Wittgenstein explicitly mentioned as an influence on him, rejected a whole tradition, comprising such exponents as Kant, Herder, Lessing and Hegel, which saw history in terms of a broadly linear progression. Spengler's *Decline of the West* portrayed history as a moving-picture of many organic-like forms that rise, unfold and subside. Each civilization is the product and final form of a culture of its own. Such cultures grow with the same aimlessness as the flowers of the field. Yet there is a uniform law of development, periodicity and decay. But this is determined by Cosmic forces which Spengler saw as lying below the level of thought. The wise man is one who yields to the 'crush of facts' and accepts that the world goes its own way, uninfluenced by what is said about it.

Such a Schopenhaurian-like pessimism is also found in Weininger, where it is tinged with a marked tendency towards solipsism:

> *The human being is alone in the cosmos, in eternal, terrible loneliness. He has no purpose outside himself, nothing outside of himself for which he lives – he has flown far from wanting to be a slave, being able to be a slave, having to be a slave: all human society recedes under him. Social ethics recedes; he is alone,* alone (From Weininger's *Sex and Character*, quoted in Rudolf Haller, *Questions on Wittgenstein*, Lincoln: University of Nebraska Press, 1988, 94).

In a later writing Weininger claimed: 'The great philosopher like the great artist possesses the whole world in himself...' (ibid.).

A further factor which contributed to the sense of the isolated self was the widespread 'doubt' about the capacity of language. To be sure, this was not any single 'doubt'. Some intellectuals doubted its capacity to deliver knowledge about the external world – a view derived from Kantian uncertainty about the 'thing-in-itself'. There were also doubts about language's capacity as a medium of communication. Karl Kraus, whose acknowledged influence on Wittgenstein has already been noted, campaigned against distortions of language. W.W. Bartley III notes that Fritz Mauthner wondered whether language could survive as a medium of communication (*Wittgenstein*, La Salle: Open Court, 1973, 51). Hofmannsthal felt that thought had been utterly stifled by concepts and that no one could any longer be sure in his own mind about what he understands. This had awakened a desperate love of those arts not dependent on language. Against this climate we can more easily appreciate – without claiming the existence of any direct influence – Wittgenstein's love of music and his view expressed in the *Tractatus* that 'There are indeed things which cannot be put into words...they make themselves manifest' (TLP 6.522).

ii. Reaction to the Predicament

Clarifying the nature of the *Weltbild* of self-concern can begin with the line of reaction to the above perceived predicament confronting the self touched on in the earlier quotation from Weininger. 'The great philosopher like the great artist possesses the whole world in himself.'

This retraction from the external world and concern with interiority, in Taylor's words with the 'nature whose impulse we feel within' (SOS 416), is the first step to the *Weltbild* of self-concern.

Such a step became almost an obsession in the late nineteenth century and onwards. Rilke writes in the seventh of the *Duino Elegies*:

> *Nirgends, Geliebte, wird Welt sein, als innen.*
> *[Nowhere, Beloved, can world exist but within]* (DE 71).

Emerson believed that we are largely strangers to ourselves and counselled a turning of attention inwards. Kraus' condemnation of politics was justified on the grounds that politics '...is what a man does in order to conceal what he is' (LWDG 17). Tolstoy defines religion as a relation of man to the world; this is significant in that what is fundamental is the *relation* and not the constitution of the world. Hofmannsthal claims that the poet's task is directed to '...the search for harmonies in himself, a harmonising of the world which he carries within himself' (HAH 11). This is a useful way of understanding his own work and its commitment to interiority.

The move to inwardness permeated European society far more broadly than the above examples might imply. Indeed, the concepts of 'artist', 'genius' and 'culture' underwent a subtle development in conjunction with the emphasis on the inward self. Alvarez provides us with a useful way of understanding the change in the idea of the artist in the later part of the last century: '...the artist was no longer responsible to polite society...his prime responsibility was towards his own consciousness...the self (had become)...the arena of the arts...' (SG 181). Genius was essentially the inner authenticity, not subject to outward standards of value, that characterised the artistic response, and – for Weininger and others – the approach of the philosopher. As Raymond Williams says in his *Culture and Society 1780–1950* (Harmondsworth: Penguin, 1963, 61), genius involved an integrity that could not be confined and had to work out its own salvation in a man. 'Culture', in at least one sense of the term, came to refer to an ideal of inner perfection and autonomy that could act as a buttress for the individual against the pressures of his turbulent milieu.

Turning more narrowly to philosophy, Schopenhauer's view of genius, with its stress on the ability to leave our own interests entirely out of sight (e.g. WWR I 185–6), is relevant here. In his thought, which the young Wittgenstein was well acquainted with, what happens on the 'outside' is less important that what happens within the subject. This idea was given its most powerful and sustained philosophical expression in Kant's ethical writings. Kant's concern with interiority was centred on the austerely ethical *motivation*. As we shall see in the section of this chapter looking at the roots of the Absoluteness-Element, the Kantian terminology crept into the Wittgenstein position but the context in which that terminology was utilised was significantly different.

Thus in what I term the *Weltbild* of self-concern, the concern with the inner life was by no means confined to an integrity solely dependent on a Kantian absence of self-interest accompanying or permeating one's motivation. It involved a seeking out of a means to buttress the self against contingencies. It involved seeking an inward sense akin to what was supposedly available through the sense of 'culture' mentioned in the last paragraph. But in Continental Europe, much more than in the Britain which is the subject of Raymond Williams' study, such a struggle led to a radical stress on individually forged resources unconnected with any of the institutions of society. Weininger advocated suicide. This solution was toyed with by Wittgenstein. It was also considered by many of Dostoyevsky's characters, an author Wittgenstein came to admire. Wittgenstein rejected suicide as the 'elementary sin' (NB 10.1.17). But, did he reject the basic motivation and analysis of the predicament of self that suicide was supposed to provide an answer to? This is something I would contest.

Consider what one of Dostoyevsky's characters, Kirilov in *The Devils*, sees suicide as the answer to:

> *Full freedom will come only when it makes no difference whether to live or not to live. That's the goal for everybody... He who conquers pain and fear will himself be a god... For three years I've been searching for the attribute of my divinity, and I've found it: ...self-will. That's all I can do to prove in the main point my defiance and my new terrible freedom. For it is very terrible. I am killing myself to show my defiance and my new terrible freedom* (F. Dostoyevsky, *The Devils*, trans. David Magarshack, Harmondsworth: Penguin, 1953, 125–126, 614–615).

Reacting to this perceived predicament, Kirilov wants to be self-sufficient, true to himself and able, in one sense, to triumph over the suffering and condemnation to annihilation he is subjected to. This is the epitomy of what I term the *Weltbild* of self concern: an obsessive seeking to give to the self a sense of independence in confronting an impersonal universe.

In the next section I shall sketch further examples in order to suggest that the Perspective-Element's idea of independence of the world is really just such a form of self-concern. Thereafter, I shall seek to show that all the other elements of the Wittgensteinian position can be understood as deriving from a world-view permeated with such self-concern.

The Perspective-Element and the *Weltbild* of Self-Concern

There are some parallels in the history of philosophy which connect with the Perspective-Element. In Plato's *Apology* there is the idea that the good man cannot be harmed. There is also the Stoics' ideal of self-mastery and Spinoza's belief that if we realize that the vicissitudes of life arise out of necessity we can be free of fear of them. Kant was hostile to a religious attitude that sought a justification for each particular unfortunate event. He wrote about a sense of the course of life as a whole:

> *(I)s it not possible to have peace and contentment, great though our wretchedness and trouble may be?...Thereby we find consolation in, though not for, the evils of life, a solid contentment with the course of life as a whole* (*Lectures on Ethics*, trans. L. Infeld, London: Methuen, 1930, 94).

However, Kant was critical of any connotations of 'independence' from the world that seeing the course of life as a whole might have. Being able to find consolation in the latter must not make the self 'independent of nature' like the Stoic who is 'like a god in the consciousness of the excellence of his person' (*Critique of Practical Reason and Other Writings on Moral Philosophy*, trans. and ed. L.W. Beck, Chicago: University of Chicago Press, 1949, 239).

The type of independence of the world that characterises the Perspective-Element is clearly found in Kierkegaard and Schopenhauer, two thinkers close in time and sentiment to Wittgenstein's *milieu*. Schopenhauer wrote in *The World as Will and as Representation* of the moral and happy man:

> *Nothing can harm him any more...for he has cut all the thousand threads of will that bind us to the world* (Quoted in C. Barrett *Wittgenstein, Ethics and Religious Belief*, Oxford: Blackwell, 1991, 54).

Schopenhauer spoke of the ability to '...leave entirely out of sight our own interest, our own willing, and our aims and consequently to discard our own personality for a time, in order to remain *pure knowing subject*, the clear eye of the world' (WWR I 186). In his essay 'On the Indestructibility of Our Essential Being by Death' he talks of the person who becomes aware of the present as the sole form of reality and aware that it

> *has its source in us, and thus arises from within and not from without... (Such a person)...cannot doubt the indestructibility of his own being...his existence will not be affected by (death)...for there has been as much reality within him as without* (*Arthur Schopenhauer: Essays and Aphorisms*, trans. R.J. Hollingdale, Harmondsworth: Penguin, 1970, 69).

Kierkegaard held that the world cannot punish an innocent man because such a person 'has the strength of eternity in resisting any infringement' (*Purity of Heart*, London: Fontana, 1961, 85). In *Fear and Trembling* he wrote of the infinite resignation which is a necessary condition of his concept of faith. (Lowrie, the translator, usefully points out that the Danish term 'resignere' is more active than the English 'resignation' and that 'renunciation' is in some contexts a better translation of it). Kierkegaard says that 'he who has made the act of resignation is infinitely sufficient unto himself. What (his lover) does cannot disturb him'. He also says that '...it is only the lower natures who find the premise of their action outside themselves' (*Fear and Trembling and The Sickness Unto Death*, trans. W. Lowrie, Princeton: Princeton University Press, 1968, 55). The latter seems to mean that in the case of those more advanced in the approach to authentic faith, the premise of their action is located firmly within themselves. Nothing in the world outside can affect that.

It is crucially important to appreciate the context of self-concern and self-absorption which is connected to the Perspective-Element's *Sitz im Leben* in Wittgenstein's view of the world. The way independence of the world involves such self-concern will now be explored in relation to Tolstoy, von Hofmannsthal, Rilke and Emerson.

Monk highlights how Wittgenstein once claimed that religious experience consisted in getting rid of anxiety (*Sorge*) and had the consequence of giving one the courage not to care what might happen (LWDG 67). This same obsession with his own anxiety permeates Tolstoy's thought. In his *Confession* Tolstoy describes how, when nearing the age of fifty, he suffered a crisis. Nothing could remove from his mind the idea that life was meaningless:

> *...there was nothing ahead other than deception of life and of happiness, and the reality of suffering and death: of complete annihilation...Today sickness and death will come...and nothing will remain...How can a person go on living and fail to perceive this?* (C 30–31).

Faced with such a sense of being trapped, he sought long and hard, as his *Confession* relates, for a sense of purpose to life that would make him immune to the sense of the threat that contingencies presented.

He finally found a form of living that gave him peace. 'We are not attracted to genuine belief by the well-being the believer is promised but by something which manifests itself as the only recourse to deliverance from all misfortune and death' (C 220). According to Tolstoy's conception of religion, religion is a means of deliverance.

How is deliverance mediated by religion? In the essay 'What is Religion and of What Does its Essence Consist?' Tolstoy says that

> *...a rational person must do integration: that is, establish a relationship to, the immediate issue of life, a relationship to the entire infinite universe in time and space, conceiving of it as a whole* (C 87).

This idea of a relationship with the *whole of life* is reminiscent of the Perspective-Element. And this link is not coincidental. Because for both what motivates, and also determines, the nature of their understanding of religion is this relationship to the whole of life and the consequent alleviation from suffering and meaninglessness that it provides. Faced with constraint, the relentless feeling of entrapment in a meaningless universe and the inability to draw sustenance from relationships with others, the solitary Tolstoyan individual yearns for a form of self-control that will provide freedom. Tolstoy had studied the Stoics in detail, especially Epictetus. He was much impressed by the latter's saying that 'A free man is only master of what he can master without impediment. And the only thing we are entirely free to master without impediment is ourselves' (C 195).

In later life Hofmannsthal turned to a pious life of adherence to the Catholic faith – he became a tertiary of the Franciscan order. By this time he had rejected his earlier preoccupation with the self's individual contemplation of life and achieved a concern for community and for other people. But for our purposes it is his earlier period that we find illustrative of a self-concern underlying the desire to see things in terms very similar to the Perspective-Element. Hammelmann gives us some indication of the nature of this when he describes Hofmannsthal' obsession with achieving a '…state (in which)…the limitations of time…seemed to be overcome; the borderline between the finite and the infinite was temporarily suspended and he felt himself freed from the burden of the here and now' (HAH 12). Hofmannsthal was preoccupied with the way the burdens of life affected his individual self and sought a state in which he might defy them.

A good example of this is to be found in his prose piece, 'The Letter of Lord Chandos', originally published in 1902, which is a fictional letter written by an Elizabethan English nobleman to Francis Bacon. This work is meant to be indicative of Hofmannsthal's own personal concerns.

Lord Chandos stresses that he is writing to lay bare his '…inner self…and to reveal a disease of my mind' (LLC 130). This disease he refers to is his loss of confidence in the power of language to convey his feelings. He hankers for the mystical awareness which he sometimes has experienced from seeing the most banal of objects.

> *In (such) moments an insignificant creature – a dog, a rat, a beetle, a crippled apple tree...mean more to me than the most beautiful, abandoned mistress of the happiest night. These mute and, on occasion, inanimate creatures rise toward me with such an abundance, such a presence of love, that my enchanted eye can find nothing in sight void of life. Everything that exists...everything touched upon by my confused thoughts has a meaning. Even my own heaviness, the general torpor of my brain, seems to acquire a meaning; I experience in and around me a blissful, never-ending interplay... (I)t is as though my body consisted of nought but ciphers, which give me the key to everything...as if we could enter into a new...relationship with the whole of existence* (LLC 137–8).

Note here the infatuation with his own feelings, with recovering a sense of the whole of existence which, in effect, nullifies the effect of individual moments of burden and frustration. Everything has to acquire a *meaning*, everything has to satisfy the isolated self's sense of individual meaningfulness. The writer here is full of his own inner sensitivities and seeks an inner, soothing sense of integrity that external constraints cannot touch.

The poet Rainer Maria Rilke's view of life also derives from a thoroughgoing self-absorption and concern to defend his own inner integrity. Graff says of him that he '...shies away from all bonds other than those which originate within him and resolve themselves in the inspired and longed-for poem' (RMRG 62). Rilke was a poet of uncompromising determination to guard his creative freedom. Graff refers to this as an 'egocentric implacability' (RMRG 63). Though, Rilke was also, to some degree, concerned that others' integrity should be safeguarded. He once described marriage as 'two solitudes' (RMRG 65). In *Letters to a Young Poet* he says that '(l)oving is basically nothing in the nature of that which we call losing oneself in....unity with another person... (but) is a noble occasion for the individual to ripen, to become something in himself, to become world, to become world for himself...' (RMRG 193).

Rilke believed that any policy aimed at the mitigation of suffering, carried out either by a state or by a charitable group, could be injurious to the suffering self's integrity. In a letter of 1924 Rilke wrote:

> *It seems to me that nothing but disorder will be established if the general endeavour (which is a delusion) presumes to attempt a schematic mitigation or removal of suffering – an attempt which encroaches on a person's freedom far more disastrously than suffering itself which...imparts to those who confide in it directions for deliverance from it, if not externally, at least internally* (RMRG 64).

Rilke is said to have been impressed by a sentence from Emerson: 'The hero is he who is immovably centred' (RMRG 280). In his life he sought to attain a state in which he was himself such a hero, such a genius. There were his periodic retreats into solitude, his distance from human entanglements, his craving for uninterrupted concentration, his disregard for private possessions and his scorn for artificial stimulants.

There is an almost religious dedication to his 'Art' in Rilke's life. Art for him is not ultimately that which manifests itself as the created object – the poem, the picture, or whatever. Rather, art for him is really the creative moment. He once wrote that if, in Thomas a Kempis' *Imitation of Christ*, the word 'art' is substituted for the word 'God' then that would take nothing away from the wisdom of the book. That creative moment, which for him is both 'Art' and 'God' is possible when we '...turn our backs on events, even on our own future, in order to throw ourselves into the abyss of our being' (*Letters to Merline*, trans. V.M. Macdonald, London: Methuen, 1951, 48–49). In this connection he talks of this 'terrible will to Art' as enabling the artist to conquer those 'monsters', the things in the world and in himself which impede his progress and threaten his autonomy. 'For in a certain sense we are at one with them; it is they, the monsters, that hold the surplus strength which is indispensable to those that must surpass themselves...

(S)uddenly we feel ourselves walking beside them…in triumph…' (49). Note here the emphasis on *triumph*, a triumph akin to the sense of self-mastery than associated with Wittgensteinian independence of the world.

Through his Art Rilke can praise the blessedness of existence. In a letter of 1918 to a friend, Rilke says that despite the joylessness he had suffered more than most, he had not lost faith in the essential goodness of life.

> *… (T)he intrication of so many fatalities and horrors...cannot confuse my judgement about the fullness and goodness and affectionateness of existence* (DE 140).

Rilke is here expressing a quite deliberate defiance. All the pains in life *cannot confuse* his judgement that, despite all these unpleasant things, what he had attained is an attitude which is able to dismiss their power to affect him.

The *Duino Elegies*, begun during a period of solitude at Duino Castle, near Trieste (the property of one of his wealthy patrons) contain a number of parallels to the Perspective-Element's stress on seeing in terms of the whole and achieving an independence of the world. In the *Elegies* Rilke seeks to invoke in us a way of seeing suffering and the contingencies of the world in a different light from that in which they are usually viewed. What repelled him, says J.B. Leishman in his commentary on the Elegies, was

> *...that half-life from which death, and all that is mysterious and inexplicable is simply excluded; that life whose consolations are provided for by conventional religion, and whose activities are the pursuit of happiness and the making of money; from which fear and mystery are banished by distraction, and where suffering is regarded as merely an unfortunate accident...* (DE 142).

In other words, Rilke is resistant to a view in which suffering is avoided. He wants to confront it with a certain sort of inner defiance which achieves a fuller state of existence. Having glimpsed this fuller state, Rilke can say

> *We wasters of sorrows!*
> *How we stare away into sad endurance beyond them,*
> *trying to forsee their end ! Whereas they*
> *are nothing else than our winter foliage,*
> *our sombre evergreen,* one
> *of the seasons of our interior year* (DE 91).

Suffering and sorrow are but one part of a *whole* of life, life that is not to be divided up. Rilke also talks of seeing the 'whole' when he says

> *...the forms of the here and now are not merely to be used in a time limited way but...instead within those superior significances in which we share...Not within a Beyond...but within a whole, within the Whole...* (DE 157).

As in the Perspective-Element, the 'meaning of the whole of life' does not reside in a 'beyond' or in something 'above' or 'behind' what is to be found in the events encountered in *this* world. For Rilke, to think in this way derives from the deficiencies of the human condition and the way it perceives itself.

In the *Elegies* the 'Angels' are one image used to portray what is wrong with the human condition. They are not used as a Christian symbol but as an example of a state of existence without the divisions that characterize the nature of human life; for example, they comprehend life and death not as opposites but as composing a single unity. Rilke wrote to his Polish translator

> *...affirmation of life* and *death reveals itself as one...We must try to achieve the greatest possible consciousness of our existence...The true form of life extends through both regions...There is neither a here nor a beyond, but only the great unity, in which the 'Angels', those beings who surpass us, are at home* (DE 110).

Significantly, the Angels' condition is one of the most acute form of independence of the world. Nothing affects them. They are immune to supplication by any other creature.

Another example in Rilke in which seeing in terms of the whole of life involves an independence of the world is to be found in *Sonnets to Orpheus*. The thirteenth sonnet of the second group begins:

> *Sei allem Abschied voran, als ware er hinter*
> *dir, wie die Winter, der eben geht.*
> *Denn unter Wintern ist einer so endlos Winter,*
> *dass, uberwinternd, dein Herz uberhaupt ubersteht.*
> *[Be ahead of all parting, as if it were behind you, like the winter that's just on its way out. For among winters, one is so endlessly winter that, having made it through the winter, your heart survives after all]* (RMRB 161).

A good explanation of this passage in its context is given by Brodsky: '...do not wait for events to catch up with you. Exercise your human will...Do not be overcome by partings, but seek them, severing ties before they are severed for you by circumstance. Do not fear growth, or solitude or death' (RMRB 161). Rilke's self-concern manifests itself in his hope that one's '...heart survives after all'; that the self's integrity will remain intact after the parting.

Note that we here have a more intense sense of independence of the world than that found in the Perspective-Element. Here, independence is not merely a matter of accommodating oneself when circumstance intrudes; it can also involve a severing of ties before circumstance intervenes. Brodsky talks of Rilke as a man '...whose whole life was a series of partings, from people he loved or who loved him, from situations that offered warmth and protection' (RMRB 161). He felt that call to break away for the sake of his dedication to Art, to the creative moment that was the distilled state of pure independence of all contingencies.

A final example of independence of the world in Rilke is to be found in his novel *The Notebook of Malte Laurids Brigge*. This lacks a conventional plot and is essentially the reaction of a young, self-absorbed, isolated, Danish poet, to life and suffering in urban Paris. Malte finally achieves some sort of ability to continue living. The nature of his stance towards the world is reflected in the parable of the Prodigal Son about whom he writes:

> *...he had detached himself from the accidents of fate...From the roots of his being there sprang the sturdy, evergreen plant of a fertile joy. He was wholly engrossed in learning to handle what constituted his inner life* (*The Notebooks of Malte Laurids Brigge*, trans. J. Linton, London: Hogarth Press, 1950, 241. Originally published in 1910).

Again, this is an inner life that secures detachment from contingencies.

A final example of the *Weltbild* of self-concern is to be found in Emerson. Emerson stressed that greatness (genius) is not the capacity to alter the world but to bring about our adaptation to it by our state of mind. Harmony with the World is attainable for us by virtue of our capacity to inwardly attain a sense of the world as a whole. Such a sense is most readily grasped through immersion in Nature.

> *There I feel that nothing can befall me in life – no disgrace, no calamity...which nature cannot repair. Standing on the bare ground, my head bathed by the blithe air, and uplifted into infinite space, all mean egotism vanishes. I become a transparent eye-ball. I am nothing. I see all* (RWECW 10).

These passages are close to Wittgenstein's idea of absolute safety and the view of things from eternity. The transparent eye-ball image is a perceptual image just like the idea of a perspective on things *sub specie aeternitatis*.

This same strain of thought is exhibited in his commendation of 'Idealism' over what he calls 'popular faith'. His idealism

> *...sees the world in God. It beholds the whole circle of persons and things, of actions and events...not as one painfully accumulated, atom after atom, act after*

> *act, in an aged creeping Past, but as one vast picture, which God paints on the instant eternity, for the contemplation of the soul. Therefore the soul holds itself off from too trivial and microscopic study of the universal tablet...It is not hot and passionate at the appearance of what it calls its own good or bad fortune...No man is its enemy. It accepts whatever befalls, as part of its lesson. It is a watcher more than a doer, and it is a doer, only that it may better watch* (RWECW 36).

In his essay 'Self-Reliance' Emerson further takes up the theme of an insight that enables 'abandonment to the nature of things'.

Emerson ends the essay 'Nature' with the stress on the need to be '...resolute to detach every object from personal relations' (RWECW 44). Just as Wittgenstein's independence of the world involves an apartness – in the sense of a lack of dependence – from people, so is such apartness also found in Emerson:

> *To be brothers, to be acquaintances – master or servant, is then a trifle or a disturbance. I am the lover of uncontained or immortal beauty* (RWECW 10).

For Emerson this requires us to '...build therefore your own world' (RWECW 45).

The pronounced individualistic nature of this is seen when we consider an entry from Emerson's Journal of 21 December 1823:

> *I consort with no species; I indulge no sympathies. I see the world, human , brute and inanimate nature; and I am in the midst of them, but not* of *them...I see cities and nations and witness passions – ...the yell of their grief...touches no cord in me; their fellowships and fashion, lusts and virtues, the words and deeds they call glory and shame – I disdain them all. I say to the Universe...thou art not my mother...Star by star, world by world, system by system shall be crushed – but I shall live.* (Quoted in Richard Geldard, *The Vision of Emerson*, Rockport, MA: Element, 1995, 11).

This defiant independence of the world is indicative of some strains in the *Weltbild* of self-concern.

Before concluding this section it is appropriate to point out that the notion of independence of the world through seeing it as a whole, which underlies *Weltbild* of self-concern, drew on earlier romanticist ideas. Indeed, we find strains of the *Weltbild* in Rousseau. In the *Reveries* the latter's rejection of externally imposed restraint becomes a desire for a kind of integrity that protects the self against external intrusions. 'Having lost all hope in this life...I learned to feed on my own substance, looking within me for all its pasturage'. As he retreated into himself on the Isle Saint Pierre, Rousseau found that the

> *...mind became like crystal, emptied of all disturbing images. The boundaries of self...became a delicate rhythm, a gentle interpretation of thought and object, seer and seen: 'What do we enjoy at such a moment? Nothing outside us, nothing if not ourselves and our own existence; as long as this feeling continues, we are self-sufficient, like God'* (Paul Zweig, *The Heresy of Self-Love*, Princeton: Princeton University Press, 1980, 164).

In this self-sufficiency the self '...sees and feels nothing but the unity of all things' (Rousseau, *Reveries of a Solitary Walker*, trans. Peter France, Harmondsworth: Penguin, 1979, 108). He is '...free and virtuous, superior to fortune and man's opinion, and independent of all circumstances' (*The Confessions of Jean-Jacques Rousseau*, trans. J.M. Cohen, Harmondsworth: Penguin, 1953, 332.). This state of being *independent of all circumstances* is attained, in Rousseau (as in Emerson and others imbued with the *Weltbild* of self-concern) by psychological, as well as physical, distance from others – shaking off the 'yoke of friendship as well as that of public opinion' (*Confessions* 338).

In conclusion, this section has sought to provide examples from Wittgenstein's 'surrounding' culture and to show how the Perspective-Element can be understood as rooted in a certain form of self-concern.

What remains to be done in the next sections of this chapter is to show how the other elements of the Wittgensteinian position are linked to and founded on such a *Weltbild* of self-concern.

The Absoluteness and Anti-Consolation Elements and the *Weltbild* of Self-Concern

There is a connection between the Absoluteness-Element, which holds that self-renouncing faith is an end in itself, and the *Weltbild* of self-concern.

Kant's emphasis on self-perfection helps us to see how this connection holds:

> *If there were no (duties to oneself) then there would be no duties whatsoever...since...I am necessitated by my own pure practical reason... (and am)...the necessitating subject in relation to myself* (*The Doctrine of Virtue*, trans. Mary J. Gregor, Philadelphia: University of Pennsylvania Press, 1964, 80).

Ethics for Kant is ultimately not what is done to others but a matter of the motivation of the self that acts. The scope of this motivation is the crucial thing.

The autonomously motivated self is that which alone can bring about the ethical. The ethical is the work of such a self, the fruit of its self-perfection. Kant says that '...it is not in so far as he is *subject* to the law that he has sublimity, but rather in so far as, in regard to this very same law, he is at the same time its *author* and is subordinated to it only on this ground' (*Groundwork of the Metaphysics of Morals*, trans. H.J. Paton and entitled *The Moral Law*, London: Hutchinson, 1948, 107).

There is a point of comparison in the way Kant has the good will as good without qualification and the *Weltbild* of self-concern has the reaction of the self as an end in itself. Both centre on the response of a *subject* rather than focus on any form of object. (In the Absoluteness-Element, what is an end-in-itself is not God but the 'believing' of the believer.)

This response of the subject is something forged by the self in a way that precludes what happens in the world affecting the genuineness of ethical value (Kant) or self-renouncing independence of the world (proponents of the *Weltbild*). The evaluation of the self's orientation to the world does not depend on what Bernard Williams calls 'moral luck', that is, on what is outside the self's control and subject to contingencies (*Moral Luck: Philosophical Papers 1973–1980*, Cambridge: Cambridge University Press, 1981, 20–39). The ethical is discontinuous with the world.

This discontinuity between the ethical and everything else connects with the division between what is an end in itself (something absolute) and what is merely conducive to some further end. The very model presupposing and incorporating such a division belongs to a conception of an absolute value in which the self's reactions are what give shape and form to the idea of 'the ethical'. *It assumes an ideal in which the self can muster the integrity and autonomy* to affirm some things as of value wholly irrespective of what they yield to the self in terms of their propensity to satisfy inclinations or desires or relative ends in the world.

The terminology of the ethical as for the sake of goodness alone was widely taken up in the nineteenth century. Kierkegaard held that the ethical

> *...reposes immanently in itself, it has nothing without itself which is its telos, but is itself telos for everything outside it...* (*Fear and Trembling and The Sickness Unto Death*, trans. and notes by W. Lowrie, Princeton: Princeton University Press, 1968, 64).

Feuerbach used similar terminology to criticise traditional Christianity with its metaphysical conception of God. In his works the Christian is accused of loving others not for their own sake but as an avenue to his own salvation. For Feuerbach, Christian love is something inherently in pursuit of relative ends: a method of winning favour from a metaphysical benefactor.

Once the idea of absoluteness is recognised as conjoined to the reactions of an autonomous subject, it is difficult – if not impossible – to introduce any meaningful *necessary* place for a metaphysical dimension to religion. This point will be left for discussion when we come to deal with the Anti-Metaphysical Element. But it is worth noting that the integrity of the self presupposed by exemplars of the *Weltbild* of self-concern, an integrity able to embrace absolute ends supposedly untouched by self-interest, was fully in keeping with a Feuerbachian rejection of orientation to a metaphysical dimension.

Hofmannsthal, Trakl, Rilke and Emerson do not, so far as I can find, utilise the *terminology* of relative versus absolute ends. Yet all, in their own way, seek an orientation to the world that is complete in itself and not a means to some further end.

In Tolstoy there is an explicit reference to the Kantian framework. In a key passage illustrating this, he begins by saying:

> *A person acts according to his faith not because he believes in things unseen, nor because he works to achieve things hoped for...but because, having defined his position in the world, it is natural for him to act according to it...* (C 97).

In other words, religion is not about factual beliefs related to quasi-empirical states of affairs nor an attempt to achieve separately specifiable ends. Rather, it is something that naturally springs from a particular relation to the world as a whole. The term 'relationship to the universe' is explained in Tolstoy's essay 'Religion and Morality (C 129–150). Tolstoy claims there are three, and only three, forms of relationship to the universe. The first consists in a person existing in the world for the purpose of attaining the greatest possible personal well-being. The second relationship involves recognizing the meaning of life not in the well-being of one individual but in that of the family, the tribe, the state or some other aggregate of people. The third relationship to the universe – the Christian, which is also found among 'the Pythagoreans, Therapeutae, Buddhists, Brahmins, Epictetus, Seneca and Marcus Aurelius and not always seen in official Christianity (C 143; C 140) – consists of self-sacrificing service to the will of God.

Tolstoy claims that every person has a relationship to the universe and therefore a religion. It is not possible to prove to a person whose life is lived within another type of relation to the world '...that he must deny himself...simply because it is necessary and worthy and a categorical imperative' (C 97). Self-renouncing faith is, for Tolstoy, akin to a Kantian categorical imperative. It arises from one form of relation to the world and is absolute in that it is not subject to assessment in terms of further ends.

Tolstoy also emphasizes how the good can only ultimately be traced to and said to exist in the individual. '(A)ll the things which are commonly considered good are worthless' (C 391). What has genuine value cannot exist as something socially valuable because people's self-interest would arise and pollute it. It would become something that was merely valued as a means to social esteem and prestige. What is of absolute value must exist within the individual and be mediated through that individual's relationship to the universe as a whole.

The Anti-Consolation Element reinforces the absoluteness of the ethical/religious orientation to the world. For if this orientation were maintained on the basis of consolation and inclination, it would cease to be an absolute. Moreover, the self's stance of independence of the world would be compromised: the self would then be tied to the world, would demand of the world to provide the realisation of its inclinations. In this we see how the *Weltbild* of self-concern requires that consolation has no place as a motivating force for religious belief. Tolstoy says: 'We are not attracted to genuine belief by the wellbeing the believer is promised but by something which manifests itself as the only recourse to deliverance' (C 220). Believing for the sake of the superficial consolation of a sense of wellbeing is inauthentic.

It would be a mistake to interpret this latter passage as implying that the believer is motivated to get something – namely *deliverance* – out of believing. For Tolstoy is not talking about a range of options from which the potential believer chooses the most attractive. He is talking about someone who comes to a particular relation with the universe, a relation in which the usual worldly standards are seen as exposed and hollow. For such an individual there is *only one* direction he can follow: the stance of an absolute acceptance of the claims deriving from his relation to the world. In other words, *deliverance* is not a matter of the most acceptable among a set of options. Rather, deliverance for Tolstoy is that absolute response which derives from a particular relation to the world.

One illustration of this can be seen from the story 'Master and Man'. It is a story of how the rich, exploitative merchant Vasili Andreevich and his meek peasant servant Nikita become lost in the snow. Gradually, as their plight becomes ever more hopeless Vasili Andreevich's entire relationship to the world changes. His motivation for undertaking the dangerous journey was the hope of making money on a good deal. The life of his servant was something superfluous to him. But in the confrontation with the inevitability of death Vasili Andreevich undergoes a change in his entire relationship to the universe. Ensuring his sick servant survives the terrifying cold becomes the meaning of his life. There is no reference to any afterlife whereby Vasili Andreevich can recoup some benefit for the sacrifice made. In accepting religion and the claim of serving his fellow man, Vasili Andreevich can not be said to seek consolation. The initial terror at the impending loss of everything he had hitherto lived for has unsettled him; it has brought him to a new relationship to the world. *But that relationship is not something sought in order to secure consolation.* Rather, any sense of peace he gets from the new relationship to the world is merely a consequence of that realignment to the world. It is *not* the motivation for that realignment.

The Unreflectiveness-Element and the *Weltbild* of Self-Concern

The idea that selflessness requires a stance of unreflectiveness is deeply linked to the *Weltbild* of self-concern and had roots in the culture surrounding it.

Schopenhauer held that it is neither necessary nor possible for an intelligent man to believe in the literal truth of religious doctrines. In contrast to knowledge involving concepts of empirical relations – which have their application in connection with seeking the satisfaction of the affirmative will – Schopenhauer posited another form of knowledge. This is a knowledge acquired through art or even through suffering, a knowledge'...that cannot be communicated, but must dawn on each of us. It therefore finds its real and adequate expression not in words, but simply and solely in deeds, in conduct, in the course of a man's life' (WWR I 370).

A further parallel to the Unreflectiveness-Element is to be found in the work of Kierkegaard. Unreflectiveness seems to be assumed in Kierkegaard's Abraham (in *Fear and Trembling*). Alastair Hannay says that this 'knight of faith' acts in 'absolute isolation' and that whatever Abraham might say in explanation of his action, '...he cannot say why he should be willing to do just this in order to prove his faith' (*Kierkegaard*, London: Routledge, 1982, 55). In other words, Abraham's faith is presented as an immediate, unreflective response totally lacking any deliberative accompaniment.

Elsewhere, Kierkegaard says that Christianity '...refuses to be understood and...the maximum of understanding which could come in questions is to understand that it cannot be understood' (*Concluding Unscientific Postscript*, trans. D. Swansen, Princeton: Princeton University Press, 1941, 191). I think the best way to understand this passage is that suggested by Phillips (FAF 275). Kierkegaard is speaking of an understanding. But it is not an understanding that something which could be understood cannot be understood. Rather, it is an understanding in the sense that we come to something in relation to which 'understanding' is not the appropriate response.

This idea is clearly paralleled in Tolstoy: 'I want to understand in such a way that everything inexplicable presents itself to me as necessarily inexplicable...' (C 78). As we saw in the last section, Tolstoy saw religion as a matter not of beliefs enshrined in language but of one's relation to the universe. The absoluteness of that relationship is seen in his insistence that philosophy and science cannot establish man's relationship to the universe.

> *Neither philosophy nor science is able to establish man's relationship to the universe, because this relationship must be established before any kind of philosophy or science can begin* (C 139).

Consider also the following:

> *A person cannot discover through any sort of movement the direction in which he ought to move...In just the same way it is impossible in philosophy to use mental effort to determine the direction in which such efforts should be made...* (C 138).

It is in the nature of movement that it is carried out in a direction; but movement is not itself the determinant of what that direction should be. (Compare Winch's statement that philosophy is unable to tell a man what he should believe just as geometry is unable to tell him where he should stand – see (EA 191).) Similarly, mental effort inevitably has a direction; but the particular course that takes cannot be used to adjudicate or judge the adequacy of mental effort. Mental effort *automatically* proceeds to establish the self's relation to the world. But no amount of linguistically enshrined analytical mental effort can assess that relation.

Not only does Tolstoy insist that it is impossible to justify through reason the superiority of one form of relation to the universe to another, he also stresses that the attempt to do so is often fraught by self-interest. The Father Sergius example discussed in Chapter 1 is one example of this. This theme also permeates the late novel *Resurrection*. The central character of that novel, Nehlyudov, sees again the woman who, through his seduction, has been brought to living as a prostitute. Tolstoy dissects in detail how both his repentance and her reaction to his offers to help her are steeped in self-interest. Maslova justified to herself why she should not respond. The view of life she had consciously adopted was that sexual intercourse was the highest good for all men, and that she had it in her power to provide, or not provide, that good. 'Maslova prized this view of life more than anything…because if she were to change her idea of life she would lose the importance it accorded her…'. This process of justifying the world-view that is most congenial to individual self-importance is carried out by virtually every character in the book.

For Tolstoy religion held as a system of *beliefs* involves *pride*. Consider the following from *The Gospel in Brief*, a work that captivated Wittgenstein when he read it during the First World War:

> *(I)t is a supreme degree of pride...to assert that a particular is a divine revelation... (N)othing more arrogant can be said than that the words spoken by me are uttered through me by God... (It is)...the avowal of oneself as in possession*

> *of the sole indubitable truth...* (reproduced in ed. W. Gareth Jones, *I Cannot Be Silent: Tolstoy's Writings on Politics, Art and Religion*, Bristol: Bristol University Press, 1989, 65).

To attain the right relation to the universe requires *oproshchatsia*, that is, making oneself simple (see E.B. Greenwood, *Tolstoy: The Comprehensive Vision*, London: Methuen, 1980, 37). 'No arguments could convince me of the truth of (the peasants') faith. Only actions…could convince me' (C 58). Being in the right relationship to the universe is what matters, not any belief or any linguistic expression of faith. One must perform *deeds* associated with religious life in order to understand. The deed precedes all linguistic expression (cf. CV #402). Any reflective linguistic attempt at understanding defeats the prospect of achieving the selflessness that is sought.

In the Lord Chandos Letter Hofmannsthal may be seen as coming close to some ideas underlying the Unreflectiveness-Element. Part of the crisis detailed in that work concerns the poets's insecurity about the power of language to convey meaning. Rilke went through a similar phase between 1899–1907; he responded by seeking to avoid as far as possible terms naming objects, trying instead to convey their 'spirit' and avoiding definitions. What is significant about Hofmannsthal is the way he connects confidence about the power of language with self-security.

Lord Chandos is saying that during the period before his crisis, he had no doubt about the ability of language to adequately name and capture the essence of objects and of the world. But he now finds that that period was one of self-orientated self-confidence:

> *…in all expression of Nature I felt myself…everywhere I was in the centre of it, never suspecting mere appearance...I felt myself the one capable of seizing…each creature (as a)…key to all the others* (LLC 132).

On this view, linguistic concepts provide a comforting sense of attachment of the self to the world. The crisis of meaning that he undergoes disrupts him from this state of self-security. In response to it he develops a self-effacing way of being in the world, 'an existence…lacking in…thought' (LLC 135), a '…mysterious, wordless, and boundless ecstasy' (LLC 139), a '…new and hopeful relationship with the whole of existence…' (LLC 138). Like Wittgenstein (TLP 6.4311; NB entry for 8.7.16), he sees this in terms of living in the present: 'the Present, the fullest, most exalted Present.'

As we have already seen, Rilke wanted to inculcate in his readers an unreflective, non-linguistic apprehension of the world around us in all its transience. The second of the *Duino Elegies* contains several points on the transience of man. As Leishman summarizes, the very identity of man 'from moment to moment is as a vapour that vanishes' (DE 111). Man is urged to reflect and to strive to achieve an attitude which will still more deeply reconcile him to the fact of '…transitoriness, into the unity of life and death and the complementariness of sorrow and joy' (DE 104). Such an attitude is achieved by attentiveness to the things and relationships that surround us. In the first of the *Elegies* these include the 'tree on a slope, to be looked at day after day', 'yesterday's walk', and the ingrained habit that has stayed with us. There is also the star waiting to be perceived, the wave that 'would rise in the past towards you', and the sound of a violin through an open window 'utterly giving itself'.

> *All this was commission.*
> *But were you equal to it? Were you not still*
> *distraught by expectancy, as though all were announcing*
> *some beloved's approach…* (DE 27).

Expectancy is incompatible with this attitude. Expectancy involves some sort of reflective awareness of possibilities, possibilities that contribute to the self's craving for continual stimulation. Rilke is directing us to cultivate an attitude without expectancy and without the self-centred reflective stance towards the world that makes expectancy possible. This implies a state very like Wittgenstein's idea of living in the present.

Emerson's transparent eye-ball image is a visual image. The stance towards the world he is concerned with is an unreflective one. The opaqueness of selfhood, which so colours the perception of the world, is removed and all can be viewed without its intrusive discolouration. Emerson was an early student of eastern philosophies. This influenced his thought profoundly and led to his moving away from the doctrinal definitions of his Unitarian background to a view of religion as a way of being orientated to the world. As Alan D. Hodder says: 'As in Buddhist philosophy, the distinctions in Emerson's essays between theology and practice simply collapsed' ('After a High Negative Way: Emerson's Self-Reliance', *Harvard Theological Review* 84, 1991, 445).

In the poem 'Kaspar Hauser Lied' Trakl strives to convey the total unreflective openness of the innocent boy found at the age of sixteen, having been locked away in total isolation and near darkness all his life. Trakl's poem is based on a novel by Wasserman and was written while Trakl had begun probationary military service as a pharmacist in the garrison hospital at Innsbruck in 1912. At times he verged onto near absolute despair, describing the city as brutal and vulgar. 'And when I think, moreover, that an alien will will perhaps cause me to suffer here for a decade, I can fall into a convulsion of tears of the most desperate hopelessness…I will…always be a poor Kaspar Hauser' (From a letter quoted in PM 123). Note here Trakl's reference to an 'alien will' and the sense of this will as external violation into the self's domain. This is close in both terminology and sense to the alien will in Wittgenstein's *Notebooks* (NB 73 ff.). Though in Trakl's case the alien will is probably something more tangible, namely, the military authorities which he, in his deteriorating psychological condition, felt wholly subjected to.

In the opening stanza we find Kaspar's uncomplicated relationship to his environment. Kasper loved the sun, the green of the wood, the singing blackbird. His '...countenance was pure. God spoke a soft flame to his heart' (PM 126). Kaspar, the child raised in the darkness, loves the light of the sun and all the beauty of the natural world which he had long been denied. In the Wasserman novel, Kaspar's tutor had been trying to introduce the concept of God to his mind. But the boy was unable to grasp anything beyond his immediate experience and replied 'Kaspar loves the sun'. Trakl begins the poem with this allusion. Kaspar's relationship to the word is totally unreflective. Therefore, it seems to follow immediately for Trakl, his countenance is *pure*. God's words of recognition find their way to his heart and not to any intellectual faculty. Trakl aspired for Kaspar's pure quality of being which can passively embrace the shadow of the murderer. His head sinks away without resistance or trace of perturbation. He is unruffled by any hint of an 'alien will'. He is the unborn one because of his lack of craving to hold onto his place in this world.

To summarise, the basic elements of the Unreflectiveness-Element are to be found among exemplars of the *Weltbild* of self-concern.

The Anti-Metaphysical Element and the *Weltbild* of Self-Concern

The Anti-Metaphysical Element stresses that self-renouncing faith involves the acceptance of all things without reference to any metaphysical dimension, either a metaphysical being or a metaphysical realm. In this section I want to show how this derives from the stress on the integrity of the self found in the *Weltbild* of self-concern.

In general terms, the compatibility can be understood in the following way. Central to the *Weltbild* of self-concern is the self's stance of, in Wittgenstein's terms, independence of the world. The self is unaffected by the course of events in the world. It has within itself the capacity for detachment from dependence on outcomes. Now it is only a very short step from this to being unaffected by anything external to the self. For one having made this step, not only are outcomes in the world inconsequential to it, but so also are outcomes and situations external to the world. External to the world here means things 'above, beyond or behind' the world, that is, things metaphysical.

As we have seen in previous sections, in Tolstoy genuine religion is not a matter of an orientation to another realm; it is a matter of one's relation to this world as a whole. The Tolstoyan self is capable of an integrity than defies the dependence of the mass of people on a metaphysical conception of religion. The roots of the Anti-Metaphysical Element are clearly found in this Tolstoyan idea.

We also find Hofmannsthal exemplifying ideas comparable to the Anti-Metaphysical Element. In the Lord Chandos letter Hofmannsthal rejects the idea of religious metaphysics:

> *Such ideas...have no power over me: they belong to the cobwebs through which my thoughts dart out into the world, while the thoughts of so many others are caught there and come to rest* (LLC 133).

Chandos is aloof from seeking the security of the 'mantle' of an orientation literally directed to a metaphysical realm. He boldly asserts that such an idea has no *power* over him.

As we have seen in previous sections, Rilke's work – especially his *Duino Elegies* – emphasises that we should accept transience and not seek any cancellation of it by way of an extension of the self's duration beyond death or by any sort of instantaneous contact with a realm over and above this one. Accepting the self's transience involves the ultimacy of the here and now, of the single moment.

In Rilke's prose piece, 'The Young Worker's Letter' there is an almost Feuerbachian rejection of the metaphysical in religion:

> *What deceit to misappropriate pictures of present delight in order to sell them behind our backs to heaven! Oh, the impoverished earth ought long ago to have called in all these loans which have been drawn on its happiness, so that the Hereafter might be adorned with them. Does death really become less opaque because these lighting devices have been dragged into place behind it? And since a vacuum cannot persist, will not all that has been taken away from earth be replaced by sham... (I)t is an insult to God (I cannot think otherwise) not to see in what is granted and permitted to us here something completely capable of making us happy to the very limit of our senses* (RMRSW 69).

The idea here is of abstracting from the beauty of the here and now, thereby diminishing the present, in order to transfer to the Hereafter. This parallels Feuerbach's claim that the diminution of this world as an offering to God is in reality an act of self-affirmation: '...the sensuality which has been renounced is unconsciously restored, in the fact that God takes the place of the material delights which have been renounced' (EC 27). Rilke is saying something not dissimilar: what is taken away from the earth is replaced by 'sham'. It is sham in the sense that ignores its glories and also because it involves debasing them through wanting to have them put elsewhere. And the integrity of self that is Rilke's ideal is able to defy all dependence on such sham.

J.B. Leishman recounts how Rilke first gave expression in his *Book of Hours* to a religion deriving from a Nietzschean rejection of otherworldliness. The '...God he so frequently invokes has no relation to the God of religion...and the prayers are addressed only to himself' ('Introduction' to *Rilke: Selected Poems*, Harmondsworth: Penguin, 1964, 12). Hollingdale gives an interpretation of Nietzsche which enables us to appreciate the nature of this connection with Rilke. Hollingdale claims that we find in Nietzsche an appeal for a

> *...new mode of transcendence (which) will have to be non-metaphysical... (This will be a)...'will to power'... (that is)...the capacity to transform the divine power over the world into power over oneself... (T)he* Übermensch *is the supreme advocate of life-affirmation through acceptance of the totality of life and...of the suffering entailed in living* (R.J. Hollingdale, 'Introduction' to *A Nietzsche Reader*, Harmondsworth: Penguin, 1977, 11).

On this view of Nietzsche, the *Übermensch* is the one that stands out and is capable of power over self: power to accept this world without dependence on a metaphysical order. This is very close to the emphasis on the integrity of the self and the aspiration for the self no longer to require the need for an external metaphysical realm which we find in the *Weltbild* of self-concern.

For the *Weltbild* of self-concern, accepting the finality of death is another fundamental sign of the self's capacity to be independent of the world and to be able to live in the present. In *War and Peace*, Pierre's realisation that there is nothing in life to be dreaded involves accepting death as a complete finality to life. That sense of finality is particularly clear in the great story 'The Death of Ivan Illych'. Illych, a proud and powerful civil servant, is struck down by terminal illness. The story records in detail his gradual decline, his fears, his aloneness, and his agonizing pain. At the end he comes to see that to his question 'Why?' '...there was no answer and could be none...'. With that realisation he has come to accept the finality of death. That same finality is also what Vasilii Andreevich in 'Master and Man' comes to accept before his death in the snow storm. In his preface to *The Gospel in Brief* (in *A Confession and What I Believe*, trans. Aylmer Maude, Oxford: Oxford University Press, 1961, 118) Tolstoy wrote: 'The true life is independent of time; it is in the present.' Like the Wittgensteinian idea of living in the present, this involves accepting the finality of death.

Emerson also displays this concern for life lived in the present rather than for any future state transcending death. As we have seen, contact with nature inspires him with the sense that '...nothing can befall me in life' (RWECW 10). In this state of absolute safety he finds his eternity in the present in which he '...beholds the whole circle of persons and things, of actions and events...as one vast picture, which God paints on the instant eternity' (RWECW 36).

As we have seen, Rilke also urges on us a radical concern with the present, rather than any future state. In the ninth of the *Duino Elegies* he can praise the earth for Death, 'that friendly Death', that 'holiest inspiration', and can proclaim:

> *Look, I am living. On what? Neither childhood nor*
> *future*
> *are growing less...Supernumerous existence*
> *wells up in my heart* (DE 89).

Immersed in the timeless present, looking neither to the past nor the future, the poet can achieve a sense of being unharmed by the vicissitudes and sufferings which threaten. This acceptance of the finality of death is also powerfully exemplified in the death of Christolph in Rilke's novel *The Notebook of Malte Laurids Brigge*. Rilke took over and developed from the Danish writer J.P. Jacobsen the notion of *der eigene Tod* (authentic death) – a dignified, anti-Christian desire to accept death as a part of life. The currency of such ideas in Wittgenstein's Vienna is most clearly indicated by the cult which sprang up after Weininger's *Sex and Character*, invigorated by Weininger's own engineered acceptance of death by an act of suicide.

It is worth briefly noting that Schopenhauer also stressed the identity of all present moments throughout time. He says of the happy man that

> *Nothing can harm him any more...for he has cut all the thousand threads of will that bind him to the world* (Quoted in C. Barrett, *Wittgenstein, Ethics and Religious Belief*, Oxford: Blackwell, 1991, 54).

In his essay 'On the Indestructibility of Our Essential Being by Death' he talks of the person who becomes aware of the present as the sole form of reality and becomes clear that it has its source in us, and thus arises from within and not from without. Such a person '...cannot doubt the indestructibility of his own being...his existence will not be affected by (death)...for there has been as much reality within him as without' (*Arthur Schopenhauer: Essays and Aphorisms*, trans. R.J. Hollingdale, Harmondsworth: Penguin, 1970, 69).

The currency of the idea of ultimate value as residing in the present rather than any future state is prevalent in von Hofmannsthal's play *Death and the Fool* where the speaker realises when close to accepting his own death that:

> *In eine Stunde kannst du Leben pressen,*
> *Mehr als das ganze Leben konnte halten*
> *[In one hour of life you can compress more life/ Than once the whole of life had space to hold]* (*Hugo von Hofmannsthal: Poems and Verse Plays*, ed. & int. M. Hamburger, London: Routledge, 1961, 135).

The idea of accepting the present is also found in Goethe, who Wittgenstein admired. For example, in the second part of *Faust* we find the claim that 'Only in the present is our happiness'. In his 'Elegie' there is the call to look straight at the moment (Augenblick), for then '…so bist du alles, bist unüberwindlich' (you will then be everything and unconquerable).

This final quotation from Goethe summarises the nature of the acceptance of the finality of death prevalent in the *Weltbild* of self-concern. Such acceptance is a route to the unconquerability of the self and is thus ultimately something self-directed in that it is a route to the self's independence of the world. The *Weltbild* of self-concern, like the Anti-Metaphysical Element, rejects the idea of God as an external agent imbued with a power which vies with worldly power. The writers enmeshed in the *Weltbild* of self-concern put forward a view of power different from the conventional view associated with controlling things.

Emerson says:

> *All power is of one kind, a sharing of the nature of the world. The mind that is parallel with the law of nature will be in the current of events, and strong with their strength…so that he is equal to whatever shall happen* (CL 62).

Power is here understood as the capacity to be 'equal to whatever shall happen'. Being strong with the strength of events is really here the ability to withstand them without being affected. That, in his view, is what power really is. So when he says: 'Life is a search after power…an element with which the world is so saturated…' (CL 59) we can understand this to mean that the very prevalence of vicissitudes in the world is really a condition for the acquisition of power (as this concept is understood by him).

Emerson's conception of power requires, for its realisation in the self, a passive stance of acceptance. He says it is not supplied by any labour but it is like the climate '…which no glass or irrigation or tillage can elsewhere rival' (CL 63). In his essay 'Spiritual Laws' a central tenet of his guidance is the admonition 'Do not choose'. That is the key to independence of events. That applies to relations with people. 'It shall be the same with my friends. I shall never woo the loveliest. I will not ask friendship or favour' (CL 262). The power he aspires to is a self-completeness that does not require supplement from anything external.

We have already seen how Tolstoy describes Pierre's acquisition of a 'personal power and strength'. In that power he finds '...an inner freedom, independent of...external circumstances' (WAP 1307). But Tolstoy is also aware that people, especially historians, frequently think of 'power' as somehow implanted in the world by strong leaders. For Tolstoy, the use of 'power' and 'genius' in connection with supposedly influential historical figures really '...do not denote anything that actually exists... (but)...indicates a certain degree of comprehension of phenomena' (WAP 1342). Moreover, it is a comprehension which fails to realise that really all of life is a 'concatenation of circumstances' (WAP 1141) which no individual can control or influence in any significant way.

Rilke also displays a similar conception of 'power'. In his prose piece 'The Young Workman's Letter' the narrator says that the power of God represents not some external force which intrudes into the world but the capacity of the self to accept and 'survive unharmed' the affliction poured on it by events in life.

In conclusion, the Anti-Metaphysical Element can clearly be seen to be comparable to the kind of defiant ideal of the self's integrity that characterises the *Weltbild* of self-concern.

Conclusions

We can conclude that the elements of the Wittgensteinian position can be understood as having their *sitz im leben* in what I have called the *Weltbild* of self-concern. This is a late nineteenth century world-view, heavily influenced by a *fin de siecle* form of neo-romanticism, in which the self is infatuated with its own inner integrity and capacity to achieve an independence from all external contingencies.

This aspiration for independence underlies the Perspective-Element in the sense that this aims for a self-mastery that defies the experience of meaninglessness and suffering. It underlies the Absoluteness-Element in that the latter implies that what has absolute value is something secured by the responses of a subject capable of generating its own integrity. In being able to do this despite all external constraints and without reference to personal consolation it also underlies the Anti-Consolation Element.

Independence of the world underlies the Unreflectiveness-Element in that it is through an unreflective view of the whole that there is the capacity to remove the self's dependence on its desires coming to fruition.

The idea of independence of the world underlies the Anti-Metaphysical Element in that it presupposes a capacity, or at least aspiration, to confront everything in life, including pain, the sense of powerlessness and the termination of the self. Moreover, to do so without reference to anything outside the self, that it, no sense of contact with or inspiration from the sense of the existence of a metaphysical reality.

We are now clearer as to the type of situation in life – the kind of view of the self's predicament and consequent self-concern – that underlies the theses of the Wittgensteinian position. The fundamentals of that position can be traced back to what Heerikhuizen refers to as the neo-romantic unease with the modern world and the seeking of '...refuge in the...egocentric dreams of happiness... (in)...a religion which held aloof from the world of action...which knew no form of worship but only solitary ecstasy...' (RMRH 33). The early Wittgenstein, on my account, fits the mould of Heerikhuizen's neo-romantic.

In the next chapter I shall turn to the significance of this finding.

4 Conclusions and Beyond

Conclusions Arising from Chapters 2–3

It is appropriate to begin this chapter by summarising what we can conclude from the discussion in this study.

In Chapter 3 we found that all the elements of the Wittgensteinian position can be understood as having their *sitz im leben* or context in life in what I have called the '*Weltbild* of self-concern'. This is a late nineteenth century world-view, heavily influenced by a form of neo-romanticism which arose in response to what was then widely conceptualised as the self's predicament. Fundamental to the *Weltbild* of self-concern is the infatuation with the self's inner integrity and capacity to achieve an *independence from all external contingencies*. This latter aspiration for independence underlies each of the theses of the Wittgensteinian position. In short, the Wittgensteinian position is inextricably bound to a world-view involving a form of *self*-concern. Even though it is couched in terms of renunciation, the real nature of that renunciation is ultimately self-directed.

It is useful at this point to recall briefly the manner in which we found in Chapter 3 that each of the Wittgensteinian elements relate directly to a self-concerned independence from the world. The *Weltbild* of self-concern underlies the Perspective-Element in the sense that this aims for a self-mastery that defies the experience of meaninglessness and suffering. It underlies the Absoluteness-Element in that the latter implies that what has absolute value is something secured by the responses of a subject capable of generating its own integrity. In being able to do this despite all external constraints and without reference to personal consolation it also underlies the Anti-Consolation Element. Independence of the world underlies the Unreflectiveness-Element in that it is through an unreflective view of the whole that there is the capacity to remove the self's dependence on its desires coming to fruition.

The idea of independence of the world underlies the Anti-Metaphysical Element in that this presupposes a capacity to live without contact with, or inspiration from the sense of the existence of, an independent metaphysical reality.

The above, therefore, summarises the fundamental relations between the Wittgensteinian position and the *Weltbild* of self-concern.

We can now succinctly state the main conclusions derivable from Chapters 2–3.

The main conclusions of Chapters 2–3 are as follows: (1) The Wittgensteinian position outlined in Chapter 1 derives from a form of neo-romanticism historically based in the nineteenth century. (2) This world-view is impregnated with a form of self-concern. (3) The case-study in Chapter 2 provides us with a distinct form of self-renunciation which has no place for the *kind of self-concern* that is integral to the form of self-renunciation which has permeated into and lies at the core of the Wittgensteinian position. When philosophers such as Phillips (CP 70) and Dilman (PPL 86 & 128) claim they are illustrating the nature of Christianity in general, we have to regard their claims as highly questionable.

The grounds for claiming that the case-study in Chapter 2 actually does represent a distinct model of self-renunciation might be questioned. In anticipation of such questioning we need to briefly state the justifications for claiming that it is distinctive and different from the Wittgensteinian position.

The following are the considerations, emanating from Chapter 2–3, which support the claim for this distinctiveness:

(1) Not one of the elements proposed as intrinsic to self-renouncing faith by the Wittgensteinian position is intrinsic in the Bernanos case-study.

(2) Not merely are these elements of the Wittgensteinian position not found to be intrinsic to self-renunciation in Bernanos, some are held to be either not inherently self-renouncing or actually incompatible with the understanding of self-renunciation. Bernanos implies that absoluteness is not inherently self-renouncing. The aspiration that one's faith be characterised by such total disinterestedness is, for Bernanos, incompatible with the humility and innocence associated with what is for him the paradigm of self-renouncing faith, namely, the spirit of childhood. Similarly, the Perspective-Element, was found not to be inherently self-renouncing in Bernanos. For Bernanos the spirit of childhood is incompatible with any desire to control the self's responses so as to purposely sever its dependence on the course of events in the world.

(3) The Bernanos model of self-renouncing faith has motifs which have no mention, sometimes no currency, in the Wittgensteinian model. The principal example here is the concept of 'the spirit of childhood'.

We can conclude, therefore, that the Bernanos case-study constitutes a conception of self-renouncing faith which is distinct from the Wittgensteinian position.

The overall conclusions can be summarised again. The Wittgensteinian position derives from a form of neo-romanticism historically based in the nineteenth century. The form of self-renunciation associated with that is impregnated with a form of *self-concern*. The Bernanos case-study provides us with an example of a distinct form of self-renunciation which has no place for such self-concern.

Implications for the Wittgensteinian Position

I now want to ask where the above conclusions leave the Wittgensteinian position. Is it irretrievably crippled or is it unaffected? I shall argue that the truth lies somewhere in between these extremes but yet raises serious considerations which deserve attention.

Let us begin by asking whether the above conclusions challenge the emphasis the Wittgensteinian position has put on the concept of self-renunciation in religious life? There is no basis for thinking this. I would wish to stress that self-renunciation is a key concept in providing a descriptive account of the sort of faith we find in Bernanos. And I think this must apply to all forms of religious belief which have any claim to reside within the Christian tradition. For it seems clear to me that in the abstract it is hard to conceive of something being *religious* in a Christian sense unless it involved some element of self-curtailment and some variety of self-renunciation. In the Christian traditions *religion* seems necessarily to involve this. Where it does not apply, what we have is not religion but superstition. The latter is usefully understood as a matter of the self aspiring to secure ends solely to satisfy its own desires. (Though this definition may not to apply to all forms of magic in other, so called 'primitive' societies.)

In this respect religion differs from morality. For it is imaginable that someone might couch morality in terms of not harming others, and of that non-harming involving no diminution of the self's claims. Now in the real world, of course, such a morality would be unsustainable. For we would soon find that not harming others all too often requires restraints to be placed on the self's own interests. However, the point is that such an abstract morality does at least seem imaginable, whereas the very idea of a *religion* as involving no self-abasement or humility or other diminution of the self's domain does not. (Again I emphasise that 'religion' here refers to the range of concepts which go under that term in the western world, a world whose sensibilities have Christian influences. As to whether this applies to what we term 'religion' in application to eastern practices is not something there is here space to discuss). Even Swinburne's somewhat crude definition of religion as a '...system which offers...salvation' does incorporate some minimal self-curtailment when he talks of a necessary requirement of it involving '...forgiveness from God and reconciliation to Him for having done what we believed morally wrong' (*Faith and Reason*, Oxford: Oxford University Press, 1981, 128).

One clear implication of this study for the Wittgensteinian position is that its implicit assumption about the uniformity of self-renunciation is no longer tenable. Self-renunciation is a complicated concept. There are distinct varieties of it. A descriptive approach in philosophy must give heed to that variety.

A defender of the Wittgensteinian position might at this point concede that more should be done to recognise this variety. *But*, it might be insisted, all the Wittgensteinian position needs to do is to accommodate this variety and things can go on as before. I now wish to argue that this is not the case. The Wittgensteinian position is in need of a more radical overhaul than that.

The reason for saying this is that in the Wittgensteinian position the concept of self-renunciation is not just one religious concept alongside others. It is not something that stands on the same 'level' as, say, prayer or immortality. In the Wittgensteinian position the concept of self-renunciation has a much more fundamental place than these concepts. Self-renunciation is the concept that supposedly *demarcates* authentic from inauthentic religion. This has one fundamental consequence.

Self-renunciation becomes *the* concept that determines how all the others are to be understood. It determines the very character of the entire model of self-renouncing faith presupposed by the Wittgensteinain position. In other words, if we have to revise the concept of self-renunciation then we have to revise the entire 'structure' which the Wittgensteinain position conceives religious belief to be.

To amplify this I shall, in the next sections, say something about how this affects the view of metaphysics in relation to religion, the nature of reflectiveness and the associated issues of questionability, doubt and commitment. After that I shall look at the implications of questioning the Wittgensteinian position's concentration on belief and language *in*, and the exclusion of belief *about*, religious practice.

i. Self-Renouncing Faith and the Metaphysical

In our examination of the Anti-Metaphysical Element in Chapter 1 the Wittgensteinian position's exclusion of the metaphysical from self-renouncing faith was highlighted. We saw that the metaphysical is conceived of as something 'elsewhere', something beneath, behind or beyond the world. The primary reason for the Wittgensteinian exclusion of the metaphysical from self-renouncing faith is the assumption that genuine self-renunciation requires the self to give up the urge for the most basic security. Dilman was highlighted as claiming:

> *...even a martyr going to his death is supported by the bond he feels unites him to God. He does not feel alone and abandoned... (His) action may be completely selfless...but it does not diminish his sense of self. It does not replace it with a void* (SILAR 115).

For Dilman – in contrast to the case of the martyr – self-renunciation *does* involve the self embracing isolation and the sense of not being bonded to anything contingent. D.Z. Phillips was cited as claiming that that extended to accepting the 'radical pointlessness in things' (RST 82). But the urge to postulate metaphysical dimensions is said to involve a failure to attain such an acceptance: 'People moved by the imperfections of this world wish that things were different' (RWE 109).

I would hope that these assumptions about self-renunciation, derived from the *Weltbild* of self-concern, can be seen to be no longer tenable as applying to self-renunciation *per se*. They apply, at most, to only one understanding of it.

The Wittgensteinian position's insistence on the necessarily anti-metaphysical character of self-renouncing faith is thus of questionable status. This is even clearer when we consider that the main vehicle for demonstrating the non-metaphysical character of religious belief by the Wittgensteinian position is what I have termed the Perspective-Element. According to the Wittgensteinian position the very idea of the perspective *sub specie aeternitatis* is directly applicable, without qualification, to religious belief. Religious belief is then, as we saw in Chapter 1, supposed to constitute looking

> *...at people and things in a way which includes the light and the dark...to see them with the whole of existence as their background; to see them* sub specie aeternitatis (RST 126).

And this involves a particular type of detachment. We saw how Dilman borrowed the following passage from Eugene O'Neill to illustrate this detachment:

> *The ecstatic moment of freedom came...the joy of belonging to a fulfilment beyond men's lousy, pitiful, greedy fears and hopes and dreams...I lost myself...I was set free...I belonged, without past and future, within a peace and unity and a wild joy, within something greater than my own life...* (PPL 126).

Here is the radical immersion in the present which we saw in Chapter 3 characterised the *Weltbild* of self-concern with its underlying aspiration for an independence from the world.

In contrast to this, we find in Bernanos' idea of the spirit of childhood an aversion to this kind of proud, detached self-concern associated with the Perspective-Element. Rather than detachment from others, the spirit of childhood exemplifies a form of renunciation in which the self actively strives for a sense of a point in things and which seeks out a sense of its bondedness to others.

All this has implications for the assessment of Wittgenstein as a religious thinker. The claim that Wittgenstein is a thoroughly religious thinker, even when he is dealing with issues of language and meaning, has been made by a number of writers in recent years. These include Norman Malcolm (WRPV), Philip R. Shields (LSLW) and H.L. Finch (*Wittgenstein*, Rockport MA: Element, 1995, 132). What is unacceptable about these claims is the uncritical use of the term 'religious'. This is particularly true of Finch, who lumps Buddhism, other eastern religions, Sufism and Christianity together and claims Wittgenstein can be understood in relation to this sort of amalgam. Shields is more clear on the issue. For him, Wittgenstein is a religious thinker in that some key elements of his thought are permeated with Augustinianism.

My own view is that Wittgenstein's thought can better be accommodated in relation to the *Weltbild* of self-concern. That, as I have argued, is importantly different from some key Christian traditions in that it is permeated with a form of self-concern that is alien to the many Christian traditions. When Cyril Barrett claims that Wittgenstein's idea of independence of the world is '...nothing more than what spiritual writers and mystics from both East and West have been saying for centuries' (*Wittgenstein on Ethics and Religious Belief*, Oxford: Blackwell, 1991, 104), we should regard this as highly questionable in relation to Christianity. It fails to take into account the way independence of the world is tied to the particular *milieu* from which Wittgenstein came – something I have sought to explore in some detail in Chapter 3 and found to be permeated with self-concern. Thus, on my view, Wittgenstein is not a religious thinker in a Christian sense because of the absence in his life and thought of an unambiguously Christian notion of self-renunciation.

Let us now turn to look at the Wittgensteinian assumption that metaphysical elements in religion necessarily involve a hankering for security. This is a highly questionable assumption, as the following points suggest. Cumulatively, if not individually, they call for a revision of the simplistic view that necessarily paints a 'metaphysical' religion as impure, self-orientated and inauthentic.

Sometimes the urge for the metaphysical arises out of compassion for others rather than security for oneself. We saw in Chapter 2 how the Priest of Ambricourt appreciates the misery of the poor in spirit and wants them to have a dream of splendour. He attacks the rich and privileged for their obsession with a metaphysical hope for an afterlife, because for them all this comes to is an unmitigated lust to retain their existing status. But for the really poor, for those entrenched in abject misery, his wish is for them to have some sustenance in the idea of a world beyond this one. A Wittgensteinian might here object and insist that what the Priest really wants is that the poor should come to see the glory of this world despite the misery that overwhelms them. However, this is really a distortion of the compassion that is at issue. The Priest's compassion reaches beyond those people capable of bearing their misery. His compassion also extends to those whose despair is too great for there to be a realistic chance of that happening.

A further problem for the Wittgensteinian position's assumption that metaphysical conceptions are incompatible with self-renunciation is the fact that sometimes the incorporation of a metaphysical dimension serves to *accentuate the degree of self-renunciation involved*. This is not a point argued for in the preceding chapters. However, further justification for the claim is to be found in my paper 'D.Z. Phillips, Self-Renunciation and the Finality of Death'. That paper provides an example of one Ignatian conception in which recognition of the indelible permanence of the self is integral to appreciating the nothingness of Man before God. The point at issue is that the Wittgensteinian position, in its blanket exclusion of metaphysical conceptions, fails to appreciate ways in which some such conceptions can be integrally linked to self-renunciation. I appreciate that much could be done by way of providing further illustrations of how this obtains. Needless to say, there is insufficient space to do that in this present study.

At this point an exponent of the Wittgensteinian position might claim that Wittgenstein's theory of meaning drives us to reject all metaphysical interpretations because it renders the idea of a metaphysical realm unintelligible. I next want to show how this line of argument is of much more limited scope than is generally assumed by exponents of the Wittgensteinian position.

Wittgenstein's view of meaning does not rule out all forms of metaphysical views within self-renouncing faith. It is true that Wittgenstein has been credited with some pretty far reaching 'metaphysics bashing'. Examples include the defusing of the idea of the Cartesian subject and the rejection of the possibility of the notion of a state of primordial certainty, a state unmediated by a subject's immersion in culture. I do not wish to challenge the legitimacy of this accreditation.

What does seem to me inappropriate is, firstly, the assumption that because some persistent metaphysical 'myths' are made suspect then all metaphysical conceptions should be taken as such. Secondly, the idea that the lack of complete intelligibility in some metaphysical conceptions should be a reason for rejecting them from the domain of self-renouncing faith.

This latter assumption ignores the possibility that the phenomenon of less than full literal transparency has a place to play in some forms of religious practice. It is important to distinguish between how words acquire meaning from the way words are subsequently used referentially in practices which, in various ways, are directed at what is taken within a practice as being literally *there*. What exists – what can intelligently be said to have existence – is often a consequence of the particular conceptualisation inherent in a practice or 'language-game'. In some practices, metaphysical entities are taken as in some ways quasi-empirical for practice-dependent reasons. In the case of religious practice such practice-dependent reasons can involve a state of self-renunciation. Before going on to try to illustrate this it is important to recognise one further distinction.

The Wittgensteinian position appears to view metaphysical beliefs in religion as of one integrated type. Yet the metaphysical is much more varied and in this connection we can offer a distinction which, though it has no claims to being absolute and comprehensive, is useful. We can distinguish between (a) the metaphysical conceived of as an ordered whole into which the relationships and regularities of this world are grafted; (b) the metaphysical as a random incursion into the secure stability of this world. Now conception (a) is holistic. All the details and particularities of this world are somehow subsumed into a grand schema. This, of course, has implications of predictability and security which the Wittgensteinian position is wary of. But conception (b) does not necessarily have these implications at all.

An example will hopefully illustrate the above distinctions and make clear how metaphysical entities can be taken as in some ways quasi-empirical for practice-dependent reasons which serve to express one mode of self-renunciation.

Henry James' writings have recently received attention for the ethical and philosophical dimension they contain. It is my view that James' writings have much to say about the theme of renunciation. In one of his works James gives us a case of renunciation in relation to the metaphysical. The young governess in *The Turn of the Screw* takes charge of two young orphaned children in a remote house and comes to believe that they are being lured into evil by the spirits of two dead servants of the same house.

Some literary critics reject the metaphysical interpretation of this work and view it as a study of the governess' delusion. However, this interpretation appears to be contrary to James' own purpose. It also overlooks the way in which a literal reading of the story gives expression to one form of the Jamesian theme of renunciation.

In the 1908 preface to the work James gives us an important insight into his view of things. He says that the

> *... essence of the matter was the villainy of motive in the evoked predatory creatures; so that the result would be ignoble – by which I mean trivial – were this element of evil but feebly or inanely suggested...*
>
> *What, in the last analysis, had I to give the sense of? Of their being, the haunting pair, capable, as the phrase is, of everything – that is of exerting, in respect to the children, the very worst action small victims so conditioned might be conceived as subject to* (Reproduced in ed. Robert Kimbrough, *The Turn of the*

> *Screw: An Authoritative Text, Background and Sources, Essays in Criticism*, New York: W.W. Norton, 1966, 122).

Thus James' purpose is to convey a certain 'villainy of motive', an acute sense of evil. It is important for the success of that purpose that a metaphysical dimension to the evil is suggested. This 'sense of horror' is missed on a psychological interpretation of the story. What needs to be conveyed is the reaction of one human being – the young governess – in the face of something terrible. Even a prominent proponent of the psychological interpretation of the work comes close to recognising it. Leon Edel writes:

> *Regardless of what any clinical diagnosis of the governess might be or any judgement of her credibility as a witness, there remains the sense of horror and the extent to which it is communicated to the reader* (*The Psychological Novel 1900–1950*, New York: J.B. Lippincott & Co, 1955, 68).

In the face of this horror the young governess' stance is one of uncompromising virtue.

After coming to believe that the spirits are out to possess the children, the young governess does not desert them. Though in fear she decides to try to become the victim instead of them.

> *I had an absolute certainty that I should see again what I had already seen, but something within me said that by offering myself bravely as the sole subject of such experience, by accepting, by inviting, by surmounting it all, I should serve as an expiatory victim and guard the tranquillity of my companions...The children...I should thus fence off and absolutely save* (TS 39).

And later:

> *I was to protect and defend the little creatures in the world the most bereaved and the most lovable. I was a screen – I was to stand before them. The more I saw, the less they would* (TS 42).

The renunciation in the above case is incomprehensible without reference to a literal belief in the metaphysical, paranormal reality of the ghosts on the part of the governess. James' story is a study in the consciousness of one who has renounced her own future in order to save the innocents. Years later, when setting out the events on paper, the female narrator writes that '...in going on with the record of what was hideous at Bly...I renew what I myself had suffered' (TS 57).

This case has an application to understanding some forms of religious belief involving the metaphysical. Firstly, such belief need not be about any wholesale metaphysical order – in this case we have transitory incursions into the present. Secondly, the belief in the incursion of metaphysical realities need not involve any precise specification of, or clarity about, the exact mechanics of the nature of the intervention. The governess has no theory about how it is possible that such entities as ghosts could exist. Nor is she in possession of any refined view of the precise nature of the evil confronting the children. James does not think it necessary or purposeful to try to explain just what this evil is. He says, 'There is for such a case no eligible absolute of the wrong' (in ed. Robert Kimbrough, *The Turn of the Screw: An Authoritative Text, Background and Sources, Essays in Criticism*, New York: W.W. Norton, 1966, 122). The particular reader has to refine further this sense of horror for himself. In the particular context in which James wrote there was a living tradition in which he might be helped. As Peter G. Beidler has shown in *Ghosts, Demons and Henry James* (Columbia: University of Missouri Press, 1989) James was writing to an audience informed by populist treatments of studies into purported supernatural phenomena. None of this should lead us to insist that James himself believed in the literal reality of ghosts as visitors from another realm. Yet it should confirm to us that James saw that such literalness was required to appreciate some forms of evil and to enable us to appreciate some forms of renunciation in confrontation with them. The phenomenon of less than full literal transparency is not a problem within a practice in which threatening ghosts are deemed to exist. There is sufficient cultural agreement on how to interpret certain types of phenomena and evidences that a fully watertight specification of just what the nature of the metaphysical realm is is not an issue. This agreement means that, as suggested earlier, words can be used referentially in that they are directed at what is taken within a practice as being literally there without having to satisfy the rigorous standards of empirical science.

This latter point is important as a potential defence against the application to this context of the type of criticism made by David Cockburn. In his fine paper 'The Supernatural' (*Religious Studies* 28, 1992, 296), Cockburn says that the idea of a 'realm of being' may '…involve the empiricist idea that a realm of which we speak is only truly independent of us if the language in which we speak of it can be thought of as being, in the end, dictated by the nature of that realm'. But in the example from James there is no place for a rigorous tracing of what 'in the end' is this realm from which the visitations come. It suffices that there is some sense of an intrusive and threatening influence, vague yet conceived of as literal.

A Wittgensteinian retort to this might be that we cannot separate the *truth* of a belief from the attitude towards the world which goes with it. However, this objection has less force in the above case than might be assumed. For use of the term 'truth' misleadingly encourages us to conceive of the matter in terms of an extra-linguistic correspondence with reality. Whereas in fact what we should focus on is the sense that *within the terms of the example* – and the sorts of practice in which it occurs in real life situations – a degree of literal ascription is present. Such focus will enable us to see that there is no *religious* confusion here.

The choice of the Jamesian example is deliberate, even though a critic might advocate that a case more closely integrated with religious belief would have been more appropriate. D.Z. Phillips, for one, is particularly keen to sever all connections between authentic religious virtue and the para-normal. Yet I wish to urge that even metaphysical conceptions involving the para-normal can have renunciatory dimensions which can relate to those found in self-renunciatory religion.

To conclude this section we can say that the Wittgensteinian position's view that authentic faith must be anti-metaphysical is open to serious question.

ii. Unreflectiveness, Unquestionability and Conceptual Subsumption

We can now consider the implications of the criticism in this essay for the Wittgensteinian position with respect to the claim of unreflectiveness, unquestionability and passivity of the self. Where does it leave the Wittgensteinian position?

Recalling the discussion in Chapter 1, the Wittgensteinian position presents religious belief as a straightforward conceptual orientation to the world. As such, it is something necessarily characterised by an unreflective immediacy. Such unreflectiveness guarantees the absence of the intrusive influence of the self, as we found in Winch's discussion of Tolstoy's Father Sergius. Reflectiveness lets in the self. It is an avenue for the self to weigh religious belief alongside other means to secure advantage and satisfaction. In Chapter 1 we also saw how D.Z. Phillips' treatment of prayers of confession yields the same stress on unreflectiveness. Faulkner's fictional character, Temple Stevens, puts her confession in articulate form – in so doing she is able to work out her own salvation. But, says Phillips, in authentic religion, prayers of confession arise precisely in situations where the self accepts there is no prospect of it gaining salvation by its own powers of articulation and reflection. In Chapter 1 we also saw how this view of religious belief as an unreflective conceptual orientation to the world is reinforced by the idea of religious belief as regulation by a *picture*. Once taken literally then that picture becomes either uncertain and ineffective or – in the case of the picture of surviving death, for example – a consolation.

In Chapter 1, at the end of the section devoted to the Unreflectiveness-Element, I used the phrase 'conceptual subsumption' to denote the nature of the passivity implied of the self by the Wittgensteinian position. The self is subsumed into a conceptual structure, a way of categorising and responding to the world. The reality of the self seems to slip out. In the words of Winch, '...the agent is this perspective' (EA 178).

It is my view that we can no longer accept that self- renouncing religious belief (as found in diverse traditions in western Christendom) is necessarily a straightforward conceptual orientation to the world. To begin with, it is far from clear whether attitudes and perspectives are as disconnected from articulation and reflectiveness as the Wittgensteinian position supposes. Are we here right to think of two sets of phenomena rigidly distinct from each other, as it implies? I would tentatively suggest that there is more scope for a degree of overlap here than is supposed. That aside, let us look deeper into the issue of what it means to portray religious belief as a conceptual orientation to the world.

It is the Wittgensteinain position's understanding of self-renunciation that promotes in that position an oversimple view of the way the religious believer is related to the world. Such a relation is supposed to consist of the self's uniform conceptual subsumption involving unreflectiveness. In some respects it is helpful to see this as a weaker form of such unreflectiveness that Jane Howarth presents in her phenomenological account of human relation to the natural world (see 'The Crisis of Ecology', *Environmental Values* 4, 1995). In opposition to what she calls modernism, a view of the isolated subject distinct from the value-free universe of discrete interacting objects linked by a chain of causality, she highlights and expounds Heidegger's phenomenological treatment of 'dwelling'. This approach seeks to unearth the pre-theoretical notion of 'dwelling' as a way of being-in-the-world prior to distinguishing between being and the world (24). This example differs from the Wittgensteinian position in that it assumes that there is a 'fundamental relatedness to the world' (24). The Wittgensteinian approach would reject the notion of such a fundamental relatedness, rightly in my view, and would view relatedness as derivative of *particular* conceptual categorisations of the world.

Yet the Wittgensteinian position is in some ways only one step removed from this phenomenological assumption. Stephen Mulhall says '...conceptual structures determine the 'essence' of reality rather than reflecting its pre-existing nature' (*On Being in the World: Wittgenstein and Heidegger on Seeing Aspects*, London: Routledge, 1990, 155). D.Z. Phillips applies this type of Wittgensteinian viewpoint in an unqualified way to *religion* when he talks of the atheist and the believer seeing different worlds (FPE 132).

This latter case is an example of the Wittgensteinian position conflating religious belief into a simple, uniformly straightforward relation between the self and the world. But does it do justice to the variety of stances that characterise the believer's position in his or her living of the religious life?

I suggest it does not. Religious life at the level of the individual is *not* one single type of orientation to the world. It is not always and invariably something determined by a set of regulating conceptual structures. Firstly, there is a place within religious life for forms of doubts which are not best described, in the terms of Winch's Father Sergius example discussed in Chapter 1, as a matter of the intrusive assertion of the self. Doubt, we saw in Chapter 2, is quite compatible with the Bernanosian concept of self-renunciation centred on the idea of 'the spirit of childhood'.

Secondly, there is a place within religious life for the concept of commitment. In support of the latter point one example of a form of commitment encountered in religious life is found in the Ignatian tradition of retreats centred on St Ignatius' *Spiritual Exercises*. I have discussed one such conception in some detail elsewhere (see SRFD). The purpose of such retreats is often to rekindle the believer's attachment to the claims of the religious life, to make those claims real and tangible in the believer's own life. Individuals go on such retreats in order to immerse their entire consciousness in meditation on the religious pictures presented in the *Spiritual Exercises*. This includes meditating on the world prior to the very existence of the self and also on various crucial points in the life of Christ, such as Gethsemane and the Crucifixion.

The point to note is that even if the individual is here trying to immerse his or her 'consciousness' in the picture, such that it regulates fully in their life, this is something *intentional*. It is not a matter of a passively orientating conceptual scheme.

The Wittgensteinian position is suspicious of the concept of intention on the basis that this could encourage the sense of the self as being able to command its responses by an act of will. But this is really an unwarranted suspicion. What we have in the Ignatian example is really a case of what we might call 'indirect intention'. The self does not *directly* intend that it will acquire the authentically self-renouncing stance. Rather it places itself in the right circumstances for the attainment of such a stance.

There is no place for the implication of unquestionability which goes with the Unreflectiveness-Element. It is in this context of purposive self-transformation that the concept of religious commitment can be seen to have a clear role. To be committed here (though not necessarily elsewhere) is to actively seek to be subsumed by (or by some part of) the Catholic framework. But such a seeking takes place within a life where the realities of that framework are less than living for the particular individual concerned.

This type of case also enables us to see that even when conceptual subsumption is achieved (through the practices of the retreat or whatever) then that is only *part* of religious belief. Self-renouncing faith is far broader than the space that can be allowed for it by modelling it as a form of conceptual subsumption. It incorporates much else besides, including doubt, uncertainty and confusion. It is therefore an oversimplification to claim that religious belief has constancy and stability of the same order as those which characterise the fundamental beliefs Wittgenstein seeks to elucidate in *On Certainty*. In this connection Phillips (RWE 151–181) and Malcolm ('The Groundlessness of Belief', in ed. S.C. Brown, *Reason and Religion*, Ithaca & London: Cornell University Press, 1977) are simply wrong (if they are generalising about religious belief).

It is important to be clear that some parts of the Wittgensteinian position can survive the above criticism. One facet that survives is the idea that religious belief *involves* (as opposed to saying it constitutes) primitive reactions. It seems to me that this idea is distinct from the idea that religious belief is a conceptual orientation to the world. Primitive reactions are somehow rooted in the psychology of the self in a way that is different from being construed as part of its conceptual framework. Primitive reactions are just ways in which we do react; reference to the idea of a conceptual framework does not have an application in explaining why we react thus. Some religious rites are grafted onto primitive reactions, or are in some way a ritualised extension of such reactions. (For example, there might be scope for portraying rites such as baptism and ordination, or the practice of kneeling at prayer, in this way). But primitive reactions seen in this way are only a very small part of religious life.

The claim that religious belief *can involve* a conceptual orientation to the world is not unacceptable. What I have taken issue with above is the unqualified assumption that this is all that self-renouncing faith is, through and through. As I have argued, not all aspects of religious belief can necessarily be taken in this way.

The idea that religious belief *can* involve *some* degree of regulation by religious pictures can also be accepted, but only if we are willing to recognise that such regulation is not the only force in a believer's life. Regulation by a picture can be compelling. But few pictures really are credible as being able to sustain *everything* that a believer has to confront in his or her living of the Christian life. Consider Wittgenstein's picture of the Last Judgement. That might motivate the believer in the sense of instilling some degree of awe into his assessment of the seriousness of the things he comes across in his day to day life. However, it is far from clear that it can serve as a focus for *all* his or her adherence to a faith that is self-renouncing. For self-renunciation also involves the kind of commitment referred to earlier; that is something other than subjection of the self out of awe. It is tenacity of purpose that survives those periods when awe is reduced to dryness of spirit.

It might be objected by a Wittgensteinian that all beliefs involve some sort of conceptual orientation to the world and that my criticism could be attacked on that basis as being un-Wittgensteinian. In reply, I would urge that the point at issue is whether self-renouncing faith is (a) solely and (b) a single, conceptual orientation to the world. Certainly various beliefs involve conceptualisations. But a whole range of such conceptualisations are present in a believer's way of interacting with the world. The weakness in the Wittgensteinian position is that it wrongly assumes that religious belief can only be self-renouncing if it is a thorough-going, single, harmonious conceptual orientation. My point is that religious belief can be self-renouncing without it being seen in this way at all. Reflectiveness, commitment, active attempts to become conceptually subsumed, doubt – all these things and more can be accommodated within self-renouncing faith. The way they are accommodated depends on understanding particular models of such faith.

A defender of the Wittgensteinian position might try to argue that the interpretation of the Unreflectiveness-Element given in Chapter 1 is invalid. D.Z. Phillips has realised that the Wittgensteinian position is open to the charge of ending up with a 'mechanistic view of human action' and of leading to a 'view of human beings as automata' (WAR 193 ff.) and is anxious to combat such a charge. In so doing he attacks the views of John Canfield, a philosopher whose writings were quoted in Chapter 1 as exemplifying the Unreflectiveness-Element.

If Phillips succeeds in his criticism of Canfield, does this not constitute a serious problem for the way I construed the Unreflectiveness-Element? It is my view that Phillips does unearth one problem with Canfield's position, a problem which constitutes a departure from Wittgenstein's own position. However, that aspect of Canfield's view given in Chapter 1 as illustrative of the Unreflectiveness-Element is unaffected by Phillips' criticism. Phillips, therefore, has provided no serious refutation of the kind of exposition of the Unreflectiveness-Element I have given. These issues will be clarified further in the following paragraphs.

Phillips criticises the way Canfield assimilates the idea of 'just doing' in Wittgenstein to that of Zen Buddhism. Canfield illustrates a Zen ideal of absence of self with the story of a boatman who is hit by an empty boat but proceeds with unconcern. Later, when he crashes into another boat carrying people, he curses and shouts. Canfield quotes the Zen story:

> *Earlier he faced emptiness, now he faces occupancy. If a man should succeed in making empty, and in that way wander through the world, then who would do him harm* (WAZ 407).

As claimed in Chapter 1, this idea of 'just doing' is a matter of an absence of articulation or emotion which are avenues for the self to enter into the arena. As such, it seems to me to be in line with a more general theme in the Wittgensteinian position which I have termed the Unreflectiveness-

Element. (As the example from Phillips in Chapter 1 shows, it is a theme found in Phillips' own work.)

But Phillips is right to point out that Canfield has, at some points, conflated this form of 'just doing' with the Zen idea of the absence of *mentalistic thinking*, and that this is an unwarranted conflation. To make this clearer let us consider Phillips' own example of a Christian belief, the belief that one should take no heed for the morrow. Taking no heed for the morrow here, says Phillips rightly, is a matter of the *character* of the kind of life a person leads. It is not a claim that the person in question lacks or has dispensed with the mentalistic concept 'thinking'.

Though Phillips' point is quite valid it does not have any implications for the kind of interpretation of the Unreflectiveness-Element I have been employing throughout this book. The Unreflectiveness-Element as I have expounded it is not to be conflated with the idea that what the Wittgensteinian position is advocating is an absence of thinking, in the way that the Zen ideal of 'just doing' involves that. The point I have stressed about the Unreflectiveness-Element is that it involves not an absence of thinking but an absence of reflective articulation as a fulcrum around which the self's interests can revolve. Thinking is not incompatible with something being unreflective in this sense. The Unreflectiveness-Element can accommodate thinking because thinking is supposedly moulded by the conceptual framework in which the subject is found.

What the above defence comes to is this. The Wittgensteinian position is still open to the charge of having a mechanistic account of human action.

Another difficulty with the Wittgensteinian position's implication of passivity of the self is worth highlighting. The passivity associated with conceptual subsumption fails to do justice to religious conceptions of *self-valuation*. Yet self-valuation is often a genuinely religious phenomenon. We can here recall the words of the Bernanos' Priest of Ambricourt after he has finally come to appreciate his own self-worth: '(I)f pride could die in us, the supreme grace would be to love oneself in all simplicity' (J 251). For Bernanos, such self-love is a pre-requisite to solidarity with others. Another form of self-valuation is found in the Ignatian view mentioned above. This self-valuation is found conjoined with a particular sense of the self's nothingness.

> *The dominion of God is immortal like myself; it begins with time and continues through eternity; death, which deprives men of all their rights, is unable to do anything against the rights of God* (*The Spiritual Exercises of St. Ignatius*, London: Burns & Oates, 1881, 26).

Any attempt to raise questions about such issues as the intelligibility of disembodied existence is likely to miss the significant sense of the contrast between God's dominion and the self's value that is here at issue. The nothingness of selfhood is here seen in the indestructible permanence of the self. And this gives rise to a sense of a special worth appertaining to the self. There is no escaping the self that one is. Not even death can free one from being what one was created to be by God in preference to '...an infinite number of creatures who were equally possible to Him, and who will forever remain in nothingness' (24). Thus, there are genuine reasons for doubting whether the implication of passivity that is implicit in the Wittgensteinian paradigm can do justice to the sense of the reality of the self and the self-value, the reverential attitude towards self, that pertains to some religious conceptions.

To summarise the above we can say that religious belief is much more complex a phenomenon than the type of uniform conceptual subsumption implied by the Wittgensteinian position. The latter fails to do justice to many religious conceptions involving commitment, the reality of the self, and of self-valuation. The ideas of an unreflective conceptual subsumption, of adherence to a 'picture' and of primitive reactions, are able to account only for a much narrower band of genuinely religious phenomena than the Wittgensteinian position appreciates.

iii. The Wittgensteinian 'Belief-In/Belief-About Distinction'

A further implication of the criticism contained in this essay as a whole is that the Wittgensteinian position's rigid distinction between belief *in* and belief *about* the religious life is untenable. This distinction is effectively a consequence of the Unreflectiveness-Element and its emphasis on what happens *in* religious practice rather than talk or belief *about* such practice. Belief-*in* demarcates all those beliefs, language uses and practices that go on in religious life; belief-*about* relates to all forms of reflective clarification and assessment of such belief in. Let us start by recalling and elaborating on examples provided in Chapter 1.

In *The Concept of Prayer* Phillips distinguishes between an account a believer may give *about* her beliefs and a *philosophical* account of such beliefs. When praying '...the believer knows what he is doing...But when he is asked to give an account of prayer...he no longer knows his way about...One is asking...for...an indication of the meaning of prayer (to give) to someone for whom prayer means very little, and often, he fails to provide an adequate one' (CP 2). Phillips says that the concept 'God' has its meaning in the role it plays in the life of believers (FPE 19). A descriptive account must be directed to understanding that role and not to what he calls 'philosophical reflection upon the reality of God'. Utterances which belong to the latter are pseudo-epistemological theories initiated in an attempt to give religion respectability in other spheres of life.

Recourse to 'role in life' does not require compatibility with what believers themselves believe about their beliefs. This is something also made particularly clear by Michael Coughlan, who offers a criterion for assessing a descriptive account . The criterion is:

> *how well (the account) can accommodate the various features which are exhibited by that aspect of life in which the concepts function, e.g. the life of prayer...What is important is which view best accords with religious practice, with how belief in God operates in the life of the believer, not with how well it compares with the believer's 'conceptual speculation'* (WPRB 236).

In short, the criterion is that of according with the religious practice.

The problem with this criterion is that it is hard to see how it can be applied if it must be taken to exclude all belief *about* religious practice by participants of that practice. A good illustration of this is to be found in Peter Winch's *The Idea of a Social Science.* Winch says '...the test of whether a man's actions are the application of a rule is not whether he can *formulate* it but whether it makes sense to distinguish between a right and a wrong way of doing things...' (ISS 58). This seems to be in line with the above criterion in that a descriptive account should heed the norms that exist in a society *rather than* the explicit self-understanding of members of that society. However, a central tenet of the book is the claim that social relations exist in and through *ideas.* It is true that Winch allows for the existence of non-discursive ideas (ISS 128), but the two examples he gives of non-discursive activity – namely 'war' (used in a further connection in (ISS 130–1)) and the glance of understanding displayed by two characters in the film *Shane* – imply a capacity by the subject to be explicitly in possession of some sort of understanding of what he or she is involved in and to have beliefs *about* that. This suggests that any strict dichotomy between belief-*in* and belief-*about* the practices and concepts occurring in social or religious life is untenable.

There is a further consideration which I suggest is overwhelmingly potent in demanding our rejection of the unqualified belief-*in*/ belief-*about* distinction. This is the fact that within religious life there are sometimes conceptions where belief-*in* requires belief-*about.* That is, to be genuinely engaged in the practice of religious belief is seen as requiring certain beliefs about the character of one's attachment to the practice and a clear self-understanding. One example is the distinction made by St John of the Cross in his *The Ascent of Mount Carmel*:

> *...many spiritual persons indulge in recreations of sense under the pretext of offering prayer...to God; and they do this in a way that must be called recreation rather than prayer, and which gives more pleasure to themselves than to*

God (Book 3, Chapter 24, in trans. & ed. E. Allinson Peers, *The Complete Works of St. John of the Cross*, London: Burns & Oates, 1935).

This notion of an approach that is more than a matter of a 'recreation of sense' is one that derives from belief-about religious practice. Yet as far as I can see it is a *religious* distinction. Indeed it has claims to be one by Wittgensteinian criteria, namely by virtue of its self-renouncing character and its anti-consolatory stance. According to St John, reflection about religious belief and practice both transforms, extends and helps determine the nature of belief in religious life. Such reflection needs to be explicit and to be centred around articulating one's attachment. It is not at all like Dilman's idea, outlined as part of the discussion of the Unreflectiveness-Element in Chapter 1, of an unarticulated evaluation of one's values. Dilman was cited as claiming that a person '…can be thoughtful and critical (without being)…a thinker. His thoughtfulness…is manifest in his life and actions… (A)lthough he is not a thinker…yet he may be receptive, open and even vulnerable to the way life tests and tries his values' (FATM 135–6). To apply this to religion can lead to a distortion of some conceptions of the religious life where reflective self-examination and explicit articulation of one's values are demanded as a prelude to an ever intensifying process of perfecting one's attachments.

Does the Wittgensteinian position really need to draw the distinction between belief-*in* and belief-*about* in the sharp way that it does in order to provide a descriptive account? There is little reason to think so. It is appropriate to refer to what Allen Janik says about the need for a descriptive approach to incorporate reflective experience as found within the practice under investigation. In a paper entitled 'Self-Deception, Naturalism and Certainty: Prolegomena to a Critical Hermeneutics' (*Inquiry* 31, 1988) Janik says:

> *Describing (the) world-picture is...getting straight about the 'grammar' or 'logic' which makes our practices coherent...It is a matter of learning to determine the meaning of expressions by looking at the way we use them. It does not depend upon taking agents' self-descriptions as the last word; for the point of words is to get a job done, not to inform others of exactly what we believe ... (T)he language in which we are instructed is not intended as a vehicle for understanding the character of the practice but for initiation into it...'* (298–9).

Janik goes on to say that in order to articulate the structure of a practice we need the '…the sort of *reflective experience* typical of coaches…' which we find in sport.

> *The coach is a person with an experienced third-person perspective on a practice, which is to say that the coach both knows how to play the sport in*

> *question and how to convey a sense of the finer points of the game to players. To know the game…is to know what a competent coach knows; for that knowledge is definitive with respect to the practice* (299).

This understanding of Wittgensteinian descriptivism has a clear place for some forms of belief-*about*.

The form of Wittgensteinian descriptivism advocated by Janik is an advance over that implied by the Wittgensteinian position as elucidated in previous paragraphs and in Chapter 1. The idea of 'coaches' in the spiritual life ties in well with the role of the Catholic writer (a role which Bernanos saw himself as trying to fulfil). The role of the 'spiritual director' in Catholic traditions also fits in here. Even Protestant traditions which place a great emphasis on the individual discerning the will of God do have a place for spiritual writers. (The 'Journal' of George Fox is one example. Another is the place given to *preachers*; their role is often to promote a transforming reflection on religious life and their sermons can end up in printed and published form).

Despite these positive aspects in Janik's elucidation of descriptivism, it too could conceivably exclude some forms of genuinely religious beliefs-*about*. For example, he says… (T)he language in which we are instructed is not intended as a vehicle for understanding the character of the practice but for initiation into it…'. Now this implies that the only acceptable beliefs-about in religion are those that emanate from those that have a claim to be 'coaches' of one sort or another. The beliefs-*about* deriving from theologians of doctrine would appear to be excluded out of hand. Certainly, many theologically inspired beliefs-about are vested in political expediencies. But descriptivism should not assume at the outset that this is always the case. Even when it is, care should be taken to investigate the nature of the expediency concerned. The incompatibility of expediency with selflessness should not be assumed at the outset. Here it is useful to recall Stuart Hampshire's point about some leaders in the Vatican having a conception of the good which requires that they sometimes have to sacrifice their own practice of humility and innocence (*Innocence and Experience*, Harmondsworth: Penguin, 1989, 174). I think that the point here is that sometimes the adoption of expediency is justified to benefit a greater good than that of the personal integrity of individual leaders. An example of such a greater good is connected with the opportunity to further the avenues for personal sanctity by many other believers in the Church.

Space does not here permit a fuller examination of the issues surrounding the place of doctrinal theology in religious life. It is sufficient to emphasise that there is likely to be more traffic between theological beliefs-*about* and the beliefs-*about* which are handled by spiritual coaches than Janik's account can allow for.

Now that we are clearer about the dispensability of a rigid belief-*in*/belief-*about* distinction, it is worth noting where it comes from. The idea that authentic religion has to do with what goes on *in* religious life and practice is one consequence of the *Weltbild* of self-concern's obsession with the instantaneous moment – what Wittgenstein calls 'living in the present'. But not all conceptions of the religious life give as much weight to the present. Some see not an instantaneous present but a *progression*. This idea of a process or progression is well illustrated in Thomas Merton's account of self-renunciation in his essay 'Pure Intention' (in his *No Man is an Island*, London: Burns & Oates, 1955). Pure intention is distinguished from impure intention. The latter '...yields to the will of God while retaining a preference for (one's) own will'. Merton stresses that there are different degrees of pure intention. There is what he calls 'right intention'. Right intention does seek to do the will of God but '...in doing so we still consider the work and ourselves apart from God and outside Him' (62). We may do a very good job but '...in doing so we will become involved in the hope of results that will satisfy ourselves'. Right intention reaches out into many plans for work done for God, plans which stand ahead of us like milestones.

> *And God is always there at the end...is always 'future'...The spiritual life of a man of right intention is always more or less provisional...more possible than actual, for he always lives as if he had to finish just one more job before he could relax and look for a little contemplation* (63–4).

Contrasting with right intention is 'simple intention'. Simple intention '...seeks and desires nothing but the supreme poverty of having nothing but God' (64). Now a man of right intention can realize that to have God and nothing else is riches enough but, says Merton, '...between the thought of such poverty and its actualization in our lives lies the desert of emptiness through which we must travel in order to find Him' (65). Again, the point to stress here is that a failure to give due attention to beliefs *about* the religious life will lead to an over-narrow view of descriptivism.

The Wittgensteinian position is sometimes accused of being prescriptive and not living up to its descriptivist claims. The above discussion shows one sense in which this charge is justified. It is justified in the sense that the Wittgensteinian position is imbued with its own set of parameters as to what authentic religion is like and it is unable to register as genuine any phenomenon that falls outside those parameters. But, as I hope I have shown, those parameters are in reality arbitrary and should not belong in a descriptive account.

A critic might view my attempts to extend the boundaries of Wittgensteinian descriptivism as tending to undermine its claim to being philosophy and reducing it to a species of piecemeal empirical investigations. Briefly in reply, philosophy requires a clear understanding of the variety of uses of words and of the difference between one practice and another. But the end is not a fascination with particularities as such (as, for example, the scholar of ancient near eastern languages is fascinated by the subtle changes between languages, or the anatomical entomologist is fascinated by the variety of insect structure). It is how such particularities impinge upon our wider thinking. Sometimes such clarity can change the way we look at the world, can wrestle us free from preconceptions and complexities that have been taken as simple truths. To that extent, a descriptive method in philosophy can, for example, incite in us a self-scrutiny conducive to helping us to free ourselves from at least some varieties of self-deception. This, in turn, links with more ancient conceptions of philosophy which view its practice as integral to appreciating our place in the world. One example of the transformation that such a descriptive approach gives rise to is the form of detachment from self-orientation found in J. Neville Ward's *The Use of Praying* (London: Epworth Press, 1967), a devotional work inspired by D.Z. Phillips' account of prayer. (The transformative effect of understanding is an example of what the continental philosopher Hans-Georg Gadamer says about the way meaning speaks to the interpreter in his or her own situation (*Truth and Method*, New York: Seabury Press, 1975, 263 ff.).)

A critic might affirm that the earlier criticism of the Wittgensteinian position's method of descriptivism for being based on the belief-*in*/belief-*about* distinction gives no clear way forward. For, it might be said, I have suggested no clear criteria by which a descriptive approach should be practised.

Two points can be made in reply. Firstly, gaining an understanding seems to me to be something that is not amenable to any precise method specifiable in advance. In the case of religion, it is something that comes through living with religious people, through reading accounts of the spiritual life and through a certain sensitivity in the observer to the pains and frustrations of the human condition. Secondly, there is at least one fruitful general direction into an understanding. This is through identifying concepts which are fundamental to governing the character of a whole host of other concepts within the religious or other tradition being investigated.

In connection with this latter point we should again call to mind how fundamental the concept of self-renunciation in Christianity is. It is not fundamental through being established by concentration on belief in the religious life; it cuts across the belief-*in*/belief-*about* 'divide'. Yet the different forms of self-renunciation give rise to different traditions of living a religious life. Other concepts, especially that of 'God' are sometimes affirmed as the most elemental in getting an understanding of Christianity.

I would urge that the concept of self-renunciation is fundamental in that it determines the nature of what can count as the self's religiously authentic relation to God. As such, it determines one's understanding of the concept 'God' itself.

Despite my criticisms of the Wittgensteinian position's over simple account of self-renunciation, it is appropriate to applaud its isolation of so fundamental a concept. That is an important contribution to the recognition of the character of religious faith as exemplified in Christianity. It is a contribution that needs to be recognised by many contemporary philosophers of religion who deal exclusively with theologically-inspired assumptions about the concept of God almost surgically isolated from the dimensions of self-renunciation which are integrally conjoined to it in the daily context of religious life.

Bibliography

One-off references are given in the text, with full bibliographic details, and are not normally included below. The following bibliography consists of works which have either been cited a number of times or have been deemed valuable as background material. For reasons of space, only a selection of such background material is included.

1. Works Relating to Chapters 1 and 4

Barrett, C., *Wittgenstein on Ethics and Religious Belief*, Oxford: Blackwell, 1991.

Beardsmore, R.W., *Moral Reasoning*, London: Routledge, 1969.

Brown, Stuart, *Do Religious Claims Make Sense*, London: SCM, 1969.

— ed., *Reason and Religion*, Ithaca & London: Cornell University Press, 1977.

— *Objectivity and Cultural Divergence*, Cambridge: Cambridge University Press, 1984 (Royal Institute of Philosophy Lectures).

Byrne, Peter, 'Arguing About the Reality of God', *Sophia* 19, 1980.

Canfield, John, 'Wittgenstein and Zen', *Philosophy* 50, 1975.

— 'Ethics Post Zen', in ed. D.Z. Phillips, *Religion and Morality*, London: Macmillan, 1995.

Cockburn, David, *Other Human Beings*, London: Macmillan 1990.

— 'The Evidence for Reincarnation', *Religious Studies* 27, 1991.

— 'The Supernatural', *Religious Studies* 28, 1992.

Coughlan, Michael, 'Wittgensteinian Philosophy and Religious Belief', *Metaphilosophy* 17, 1986.

Dilman, Ilham, 'Wittgenstein on the Soul', in ed. G. Vesey, *Understanding Wittgenstein*, London: Macmillan, 1974.

— *Studies in Language and Reason*, London: Macmillan, 1981.

— *Freud and the Mind*, Oxford: Blackwell, 1984.

— 'Reason, Passion and the Will', *Philosophy* 59, 1984.

— *Love and Human Separateness*, Oxford: Blackwell, 1989.

— 'Self-Knowledge and the Possibility of Change', in ed. Mary I. Bockover, *Rules, Rituals and Responsibility: Essays Dedicated to Herbert Fingarette*, La Salle: Open Court, 1991.

— *Philosophy and the Philosophic Life*, London: Macmillan, 1992.

Finch, Henry Le Roy, *Wittgenstein*, Rockport MA: Element, 1995.

Gaita, Raimond, *Good and Evil: An Absolute Conception*, London: Macmillan, 1991.

Helm, Paul, *Varieties of Belief*, London: Allen & Unwin, 1973.

— 'The Perfect and the Particular', Inaugural Lecture, King's College London, June 1994.

Hertzberg, Lars, 'Language, Philosophy and Natural History' in his *The Limits of Experience*, Helsinki: Acta Philosophica Fennica, 1994.
Holland, R.F., 'Absolute Ethics, Mathematics and the Impossibility of Politics', in ed. G. Vesey, *Human Values*, Sussex: Harvester Press, 1978.
— *Against Empiricism*, Oxford: Blackwell, 1980.
James, Henry, *The Turn of the Screw and Other Stories*, Harmondsworth: Penguin, 1969.
Janik, Allan, 'Self-Deception, Naturalism and Certainty: Prolegomena to a Critical Hermeneutics', *Inquiry* 31, 1988.
Johnston, Paul, *Wittgenstein and Moral Philosophy*, London: Routledge, 1989.
— *Wittgenstein: Rethinking the Inner*, London: Routledge, 1993.
Kerr, Fergus, *Theology After Wittgenstein*, Oxford: Blackwell, 1986.
Malcolm, Norman, 'Anselm's Ontological Argument', *Philosophical Review* 69, 1960.
— 'The Groundlessness of Religious Belief', in ed. Stuart C. Brown, *Reason and Religion*, Ithaca & London: Cornell University Press, 1977.
— *Wittgenstein: A Religious Point of View*, London: Routledge, 1993.
McGhee, Michael, ed., *Philosophy and the Spiritual Life*, Cambridge: Cambridge University Press, 1992 (Royal Institute of Philosophy Supplement 32).
— 'The Locations of the Soul', *Religious Studies* 32, 1996.
Monk, Ray, *Ludwig Wittgenstein: The Duty of Genius*, London: Vintage, 1991.
Mulhall, Stephen, *Faith and Reason*, London: Duckworth, 1994.
Phillips, D.Z., *The Concept of Prayer*, London: Routledge, 1965.
— *Death and Immortality*, London: Macmillan, 1970.
— *Faith and Philosophical Enquiry*, London: Routledge, 1970.
— 'Religious Belief and Language-Games', in ed. Basil Mitchell, *The Philosophy of Religion*, Oxford: Oxford University Press, 1971.
— *Religion Without Explanation*, Oxford: Blackwell, 1976.
— 'Ffydd Athronydd - Eglurdeb neu Atebion?', *Efrydiau Athronyddol* 46, 1983.
— *R.S. Thomas: Poet of the Hidden God*, London: Macmillan, 1986.
— *Faith After Foundationalism*, London: Routledge, 1988.
— 'My Neighbour and My Neighbours', *Philosophical Investigations* 12, 1989.
— *Interventions in Ethics*, London: Macmillan, 1992.
— *Wittgenstein and Religion*, London: Macmillan, 1993.
Rhees, Rush, *Without Answers*, London: Routledge, 1970.
Shields, Philip R., *Logic and Sin in the Writings of Ludwig Wittgenstein*, Chicago: Chicago University Press, 1993.
Shooman, A.P., *The Metaphysics of Religious Belief*, Aldershot: Avebury, 1990.
Sutherland, Stewart, *Atheism and the Rejection of God*, Oxford: Blackwell, 1977.
— *God, Jesus and Belief*, Oxford: Blackwell, 1984.
Taylor, Charles, *Sources of the Self: The Making of the Modern Identity*, Cambridge: Cambridge University Press, 1989.
Tessin, Timothy, & von der Ruhr, Mario, eds., *Philosophy and the Grammar of Religious Belief*, London: Macmillan, 1995.
Thomas, Emyr Vaughan, 'D.Z. Phillips, Self-Renunciation and the Finality of Death', *Religious Studies* 28, 1992.

Tilghman, B.R., *Wittgenstein, Ethics and Aesthetics*, London: Macmillan, 1991.
— *An Introduction to the Philosophy of Religion*, Oxford: Blackwell, 1994.
Winch, Peter, *The Idea of a Social Science and it Relation to Philosophy*, London: Routledge, 1958.
— 'Understanding a Primitive Society', *American Philosophical Quarterly* 1, 1964.
— *Ethics and Action*, London: Routledge, 1972.
— *Trying to Make Sense*, Oxford: Blackwell, 1987.
— *Simone Weil: The Just Balance*, Cambridge: Cambridge University Press, 1989.
Wisdom, John, 'Gods', in his *Philosophy and Psychoanalysis*, Oxford: Blackwell, 1952.
Wittgenstein, Ludwig, *Philosopical Investigations*, eds. G.E.M. Anscombe and R. Rhees, trans. G.E.M. Anscombe, Oxford; Blackwell, 1953.
— *Tractatus Logico-Philosophicus*, trans. D.F. Pears and B.F. McGuinness, London: Routledge, 1961.
— *Notebooks 1914-16*, trans. G.E.M. Anscombe and G.H. von Wright, Oxford: Blackwell, 1961.
— 'Lecture on Religious Belief', in *Lectures and Conversations on Aesthetics, Psychology and Religious Belief*, ed. C. Barrett, Oxford: Blackwell, 1966.
— 'Lecture on Ethics', *Philosophical Review* 74, 1968.
— *On Certainty*, eds. G.E.M. Anscombe and G.H. von Wright, trans. D. Paul and G.E.M. Anscombe, Oxford: Blackwell, 1969.
— 'Remarks on Frazer's "Golden Bough"', in ed. C.G. Luckhardt, *Wittgenstein: Sources and Perspectives*, Sussex: Harvester, 1979.
— *Culture and Value*, eds. G.H. von Wright and H. Nyman, trans. P. Winch, Oxford: Blackwell, 1980.

2. Works Relating to Chapter 2

Beaumont, Ernest, 'Georges Bernanos, 1888-1948', in ed. J. Cruickshank, *The Novelist as Philosopher*, London: Oxford University Press, 1962.
— 'The Supernatural in Dostoevsky and Bernanos: A Reply to Professor Sonnenfeld', *French Studies* 23, 1969.
Bernanos, Georges, *Sous le soleil de Satan* (1926) trans. Pamela Morris, *The Star of Satan*, London: The Bodley Head, 1940.
— *La Joie*, Paris: Plon, 1929.
— *Jeanne – relapse et sainte*, Paris: Plon, 1934.
— *Journal d'un cure de campagne* (1936), trans. Pamela Morris, *Diary of a Country Priest*, London: Fontana, 1956.
— *Nouvelle Histoire de Mouchette* (1937), trans. J.C. Whitehouse, *Mouchette*, London: The Bodley Head, 1966.
— *Dialogues des Carmelites* (1949), trans. Gerard Hopkins, *The Carmelites*, London: Fontana, 1961.
Flower, John, 'The Comtesse Episode in the "Journal d'un cure de campagne"', *French Review* 42, 1962.
Hebblethwaite, Peter, *Bernanos*, London: Bowes & Bowes, 1965.

Sonnenfeld, Albert, 'The Catholic Novelist and the Supernatural', *French Studies* 22, 1968.
Speaight, Robert, *Bernanos: A Study of the Man and the Writer*, London: Harvill Press, 1973.
Whitehouse, J.C., 'A Certain Idea of Man: The Human Person in the Novels of Georges Bernanos', *Modern Language Review* 80, 1985.
— 'Teaching, Witness and Vision: Some Reflection on Bernanos' View of the Responsibilities of the Catholic Writer', *Romantic Review* 11, 1987.

3. Works Relating to Chapter 3

Alvarez, A., *The Savage God: A Study of Suicide*, London: Weidenfeld & Nicholson, 1971.
Bartley, W.W., *Wittgenstein*, La Salle: Open Court, 1985.
Bouveresse, J., '"The Darkness of This Time": Wittgenstein and the Modern World', in ed. A. Phillips Griffiths, *Wittgenstein: Centenary Essays*, Cambridge: Cambridge University Press, 1991.
Brodsky, Patricia Pollock, *Rainer Maria Rilke*, Boston: Twayne Publishers, 1988.
Emerson, Ralph Waldo, *The Collected Works of Ralph Waldo Emerson Volume 1: Nature, Addresses and Lectures*, introduction and notes by R.E. Spiller, Cambridge, MA: Belkap Press of Harvard University Press, 1971.
— *The Conduct of Life*, Edinburgh: Otto Schulze & Company, undated (First published 1860).
Feuerbach, L., *The Essence of Christianity*, New York: Harper & Rowe, 1957.
Gatzke, K., 'Wittgenstein's Solipsistic Ethics: On Trying to Tell Gretl', in eds. S. Teghrarian, A. Serafini and E.M. Cook, *Ludwig Wittgenstein: A Symposium on the Centennial of His Birth*, New Hampshire: Longwood Academic, 1990.
Geldard, R., *The Vision of Emerson*, Rockport, MA: Element, 1995.
Graff, W.L., *Rainer Maria Rilke:Creative Anguish of a Modern Poet*, Princeton: Princeton University Press, 1956.
Haller, Rudolf, *Questions on Wittgenstein*, Lincoln: University of Nebraska Press, 1988.
Hamburger, M., ed., *Hugo von Hofmannsthal Poems and Verse Plays*, London: Routledge, 1962.
— ed., *Georg Trakl: A Profile*, Manchester: Carcanet Press, 1984.
Hammelmann, H.A., *Hugo von Hofmannsthal*, London: Bowes & Bowes, 1957.
Heerikhuizen, F.W. van, *Rainer Maria Rilke*, trans. F.G. Renier and Anne Cliff, London: Routledge, 1951.
Heller, Erich, *The Disinherited Mind: Essays in Modern German Literature and Thought*, London: Bowes and Bowes, 1975.
Hofmannsthal, Hugo von, 'The Letter of Lord Chandos', in *Hugo von Hofmannsthal: Selected Prose*, trans. M. Hottinger & T. J. Stern, London: Routledge, 1952.
Jacobsen, David, 'Vision's Imperative: "Self-Reliance" and the Command to See Things As They Are', *Studies in Romanticism* 29, 1990.

Janik, Allan and Toulmin, Stephen, *Wittgenstein's Vienna*, London: Simon and Schuster, 1973.

— 'Wittgenstein, Ficker and "Der Brenner"', in ed. C.G. Luckhardt, *Wittgenstein: Sources and Perspectives*, Sussex: Harvester Press, 1979.

Kenny, Anthony, *The Legacy of Wittgenstein*, Oxford: Blackwell, 1984.

McGuinness, Brian, 'The Mysticism of the "Tractatus"', *Philosophical Review* 75, 1966.

— ed., *Wittgenstein and His Times*, Oxford: Blackwell, 1982.

— *Wittgenstein: A Life. Young Ludwig 1889-1921*, Berkeley: University of California Press, 1988.

Phillips Griffiths, A., 'Wittgenstein, Schopenhauer and Ethics', in ed. G. Vesey, *Understanding Wittgenstein*, London: Macmillan, 1974.

Redpath, Theodore, *Ludwig Wittgenstein: A Student's Memoir*, London: Duckworth, 1990.

Rilke, Rainer Maria, *The Notebook of Malte Laurids Brigge*, trans. John Linton, London: Hogarth Press, 1954.

— 'Young Workman's Letter', from *Rainer Maria Rilke: Selected Works*, trans. G. Craig Houston, London: Hogarth Press, 1957.

— *Duino Elegies*, trans. by J.B. Leishman and Stephen Spender, London: Chatto & Windus, 1975.

Schopenhauer, Arthur, *The World As Will and As Representation*, trans. E.F.J. Payne, New York: Dover, 1969 (2 volumes).

Sharp, F.M., *The Poet's Madness: A Reading of Georg Trakl*, Ithaca & London: Cornell University Press, 1981.

Tolstoy, Leo, *War and Peace*, trans. Rosemary Edmonds, Harmondsworth: Penguin, 1957.

'The Gospel in Brief', in *A Confession and What I Believe*, trans. Aylmer Maude, Oxford: Oxford University Press, 1961.

— *Resurrection*, trans. Rosemary Edmonds, Harmondsworth: Penguin, 1966.

— 'Master and Man', in *The Raid and Other Stories*, trans. Louise & Aylmer Maude, Oxford: Oxford University Press, 1982.

— 'The Death of Ivan Illych', in *The Raid and Other Stories*, trans. Louise & Aylmer Maude, Oxford: Oxford University Press, 1982.

— *A Confession and Other Religious Writings*, trans. Jane Kentish, Harmondsworth: Penguin, 1987.